ADVANCES IN HURRICANE RESEARCH - MODELLING, METEOROLOGY, PREPAREDNESS AND IMPACTS

Edited by **Kieran Hickey**

Advances in Hurricane Research - Modelling, Meteorology, Preparedness and Impacts
http://dx.doi.org/10.5772/3399
Edited by Kieran Hickey

Contributors

Eric Hendricks, Melinda Peng, Alexander Grankov, Vladimir Krapivin, Svyatoslav Marechek, Mariya Marechek, Alexander Mil`shin, Evgenii Novichikhin, Sergey Golovachev, Nadezda Shelobanova, Anatolii Shutko, Gary Moynihan, Daniel Fonseca, Robert Gensure, Jeff Novak, Ariel Szogi, Ken Stone, Xuefeng Chu, Don Watts, Mel Johnson, Gunnar Schade, Qin Chen, Kelin Hu, Patrick FitzPatrick, Dongxiao Wang, Kieran Richard Hickey

Published by InTech
Janeza Trdine 9, 51000 Rijeka, Croatia

Publishing Process Manager Iva Lipovic
Typesetting InTech DTP Team
Cover InTech Design Team

Additional hard copies can be obtained from orders@intechopen.com

Advances in Hurricane Research - Modelling, Meteorology, Preparedness and Impacts,
Edited by Kieran Hickey
p. cm.
ISBN 978-953-51-0867-2

For EU product safety concerns:
IN TECH d.o.o., Prolaz Marije Krucifikse Kozulić 3, 51000 Rijeka, Croatia,
info@intechopen.com or visit our website at intechopen.com.

Meet the editor

Dr. Kieran R. Hickey is a Lecturer in Physical Geography in the School of Geography and Archaeology in the National University of Ireland Galway, Republic of Ireland. He earned his B.A. in Geography and Economics in 1986 and his M.A. in Geography in 1990 from University College Cork, Republic of Ireland and his D.Phil from Coventry University, England in 1997. His research is in storms and hurricanes, climate change, historical climatology and climate disasters. He is a Fellow of the Royal Meteorology Society and the Royal Geographical Society and is currently hurricane editor for the International Journal of Meteorology. He has published extensively in many academic journals, edited volumes and written three books.

Contents

Preface

Although extensive research has been carried out on tropical cyclones, there is still much more to be done in order to understand them. This includes how they form, develop and move, their predictability, their meteorological signatures and their impacts, along with issues of how different societies prepare and manage or in many cases fail to manage the risk when tropical cyclones make contact with human societies.

The recent effects of Hurricane Sandy /Tropical Storm Sandy in 2012 emphasises these issues especially in the context of the vulnerability of different communities to the catastrophic impacts of these events whether in a developing country or developed urban areas such as New Jersey and New York. It is estimated that over 200 people have died in the USA, Haiti, Cuba and other countries and the cost of Sandy will be well in excess of $52 billion, of this figure at least $50 billion will be the cost of the damage done in the USA alone. But we must not forget that tropical cyclones are a devastating global phenomenon with major events affecting many parts of the world on an annual basis. For example, in 2012 the NW Pacific typhoon season has been very active, generating over 500 fatalities and around $4.4 billion dollars in damage , affecting many countries in this region.

This book provides a wealth of new information, ideas and analysis on some of the key unknowns in hurricane research at present including modelling, predictability, the meteorological footprint of cyclones, the issue of evacuation, impact of event on nutrient movement during hurricane-induced high stream flow events, the critical provision of children's medical services and the general impact of events. The book is divided into three parts and each part is organized by topic. Each part in turn is organised as logically as possible.

The first part of the book is concerned with a number of aspects of the modelling of tropical cyclones. The first chapter reviews numerical prediction systems for tropical cyclone development and the strengths and weaknesses of each of the three major approaches are identified. The second chapter in this section assesses the use of remote sensing methods for tropical cyclone development in oceans. Two case studies are considered, that of Hurricane Katrina in 2005 and Hurricane Humberto in 2007. The final chapter here assesses a parametric surface wind model for tropical cyclones in the Gulf of Mexico and in particular focussing on ten hurricanes which affected this region between 2002 and 2008, starting with Hurricane Isidore and finishing with Hurricane Ike, and again, including Hurricane Katrina.

The second part of the book examines the meteorological context of tropical cyclones. The first chapter here presents a detailed micrometeorological analysis of the wind as Hurricane Ike passed over Houston, Texas in 2008. Temperature, pressure and humidity were also incorporated into the analysis. The second chapter in this section analyses the meteorological passage of 52 tropical cyclones as they pass over part of the South China Sea, a particular focus being on wind fields, air temperature and heavy rainfall.

The third part of the book focuses on the preparation for and impact of tropical cyclones in a number of contexts. The first chapter uses simulation modelling in order to evaluate evacuations by motorised vehicles in Alabama and this has significant implications for not just the USA but also all vulnerable areas with a high usage of motor vehicles. The second chapter looks at the influence of high stream-flow events in the post hurricane period and the direct effect of this on nutrient flows into wetlands, in particular the focus is on nitrate and ammonium flows. The third chapter in this section reviews the medical needs, both physical and psychological of children in a post hurricane scenario. Much of this research having being carried out as a result of the impact of Hurricane Katrina in the USA and in particular the need for systematic intervention is identified in the case of psychological health problems being presented by individual children. The final chapter assesses the meteorological and human impact of both Hurricanes Debbie and Charley on Ireland but also with reference to the UK and Europe. Both caused significant damage and loss of life but were very different in character, Hurricane Debbie bringing record high winds to Ireland and Hurricane Charley bringing record rainfall to Ireland and consequently severe flooding in some locations.

Kieran R. Hickey
School of Geography and Archaeology
AC125, Arts Concourse Building
National University of Ireland Galway
Galway City, Republic of Ireland

Section 1

Modelling

Chapter 1

Initialization of Tropical Cyclones in Numerical Prediction Systems

Eric A. Hendricks and Melinda S. Peng

Additional information is available at the end of the chapter

http://dx.doi.org/10.5772/51177

1. Introduction

Tropical cyclones (here after TCs) are intense atmospheric vortices that form over warm ocean waters. Strong TCs (called hurricanes in the North Atlantic basin, or typhoons in the western north Pacific basin) can cause significant loss of lives and property when making landfall due to destructive winds, torrential rainfall, and powerful storm surges. In order to warn people of hazards from incoming TCs, forecasters must make predictions of the future position and intensity of the TC. In order to make these forecasts, a forecaster uses a wide suite of tools ranging from his or her subjective assessment of the situation based on experience, the climatology and persistence characteristics of the storm, and most importantly, *models*, which make a prediction of the future state of the atmosphere given the current state. In this chapter, the focus is on dynamical models. A dynamical model is based on the governing laws of the system, which for the atmosphere are the conservation of momentum, mass, and energy. Since the system of partial differential equations that govern the atmosphere is highly nonlinear, a numerical approximation must be made in order to obtain a solution to these equations. Short term (less than 7 days) numerical weather prediction is largely an initial value problem. Therefore it is critical to accurately specify the initial condition. The accuracy of the initial condition depends on the forecast model itself, the quality and density of observations, and how to distribute the information from the observations to the model grid points (data assimilation). Since most TCs exist in the open oceans, most observations come from satellites, and often intensity and structure characteristics are inferred from the remotely sensed data [10]. Therefore a key problem that remains for TC initialization is the lack of observations, especially in the inner-core (less than 150 km from the TC center).

TCs are predicted using both global and regional numerical prediction models. Global models simulate the atmospheric state variables on the sphere, while regional model simulate the variables in a specific region, and thus have lateral boundaries. Due to smaller domains of interest, regional models can generally be run at much higher horizontal resolution than global models, and thus they are more useful for predicting tropical cyclone intensity and structure. As an example of how well TC track and intensity has historically been predicted, Fig. 1 shows the average track and intensity errors from official forecasts from the National Hurricane Center from 1990-2009. While there has been a steady improvement in the ability to predict track (left panel), there has been little to no improvement in this time period in the prediction of TC intensity (right panel). Currently there is a large effort to improve intensity forecasts: the National Oceanic and Atmospheric Administration (NOAA) Hurricane Forecast Improvement Project (HFIP).

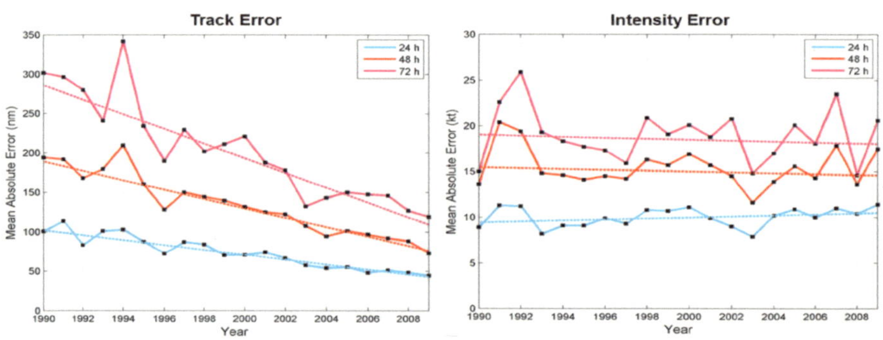

Figure. 1. Average mean absolute errors for official TC track (left panel) and intensity (right panel) predictions at various lead times in the North Atlantic basin from 1990-2009. Data is courtesy of the National Hurricane Center in Miami, FL, and plot is courtesy of Jon Moskaitis, Naval Research Laboratory, Monterey, CA.

Errors in the future prediction of TC track, intensity and structure in numerical prediction systems arise from imperfect initial conditions, the numerical discretization and approximation to the continuous equations, model physical parameterizations (radiation, cumulus, microphysics, boundary layer, and mixing), and limits of predictability. While improvements in numerical models should be directed at all of these aspects, in this chapter we are focused on the initial condition. The purpose of TC initialization is to give the numerical prediction system the best estimate of the observed TC structure and intensity while ensuring both vortex dynamic and thermodynamic balances. In this chapter, a review of different types of TC initialization methods for numerical prediction systems is presented. An overview of the general TC structure and challenges of initialization is given in the next section. In section 3, the direct vortex insertion schemes are discussed. In section 4, TC initialization methods using variational and ensemble data assimilation systems are discussed. In section 5, initialization schemes that are designed for improved initial balance are discussed. A summary is provided in section 6.

2. Overview of the TC structure

Tropical cyclones come in a wide variety of different structures and intensities. Intensity is a measure of the strength of the TC, and is usually given in terms of a maximum sustained surface wind or the minimum central pressure. Structure is a measure of various axisymmetric and asymmetric features of the TC in three dimensions. Structure encompasses the outer wind structure (such as the radius of 34 kt wind), inner core structure (such as the radius of maximum winds, eyewall width and eye width), as well as various asymmetric features (inner and outer spiral rain bands, asymmetries in the eyewall, asymmetric deep convection, and asymmetries due to storm motion and vertical wind shear). Additionally, structure would encompass vertical variations in the TC (such as the location of the warm core and how fast the tangential winds decay with height). While there are some observations (particularly for horizontal aspects of the structure from remote satellite imagery), there are never enough observations to know the complete three-dimensional flow and mass field in the TC.

In this section we outline some important structural aspects of the TC, including the basic axisymmetric and asymmetric structures that should be incorporated into the numerical model initial condition. An atmospheric state variable ψ, which may be temperature or velocity, may be interpolated to a polar coordinate system about the TC center and decomposed as $\psi(r, \phi, p, t)=\overline{\psi}(r, p, t)+\psi'(r, \phi, p, t)$, where $\overline{\psi}(r, p, t)$ is the axisymmetric component of the variable (where the overbar denotes as azimuthal mean), and $\psi'(r, \phi, p, t)$ is the asymmetric component of the variable. Here r is the radius from the vortex center, ϕ is the azimuthal angle, p is the pressure height, and t is the time. Often TCs are observed to be mostly axisymmetric (but with lower azimuthal wavenumber asymmetries due to storm motion and vertical shear), however in certain instances, and in certain regions of the TC, there can be large amplitude asymmetric components.

2.1. Axisymmetric structure

Fig. 2 shows the basic axisymmetric structure of a TC from a real case, Hurricane Bill (2009), obtained from the initial condition of (COAMPS®) numerical prediciton system [1] shown. In the Fig. 2a, the azimuthal mean tangential velocity is shown, in Fig. 2b the radial velocity is shown, and in Fig. 2c the perturbation temperature is shown. There are three important regimes in Fig. 2: (i) the boundary layer, (ii) the quasi-balance layer, and (iii) the outflow layer. The boundary layer is the region of strong radial inflow near the surface in Fig. 2b. Above the boundary layer, the winds are mostly tangential in the quasi-balance layer, and then at upper levels (Fig. 2b) the outflow layer with strong divergence and radial outflow is evident. In Fig. 2a, it can be seen that the strongest tangential winds are near the surface and decay with height, and in Fig. 2c a mid to upper level warm core is evident. While this is just one case, it illustrates the basic axisymmetric structure of a TC. While the vertical velocity is not shown in this figure, there exists upward motion in

1 COAMPS® is a registered trademark of the Naval Research Laboratory

the eyewall region, and this combined with the low to mid-level radial inflow and upper level outflow constitute the hurricane's secondary (or transverse) circulation. Changes in the secondary circulation are largely responsible for TC intensity change.

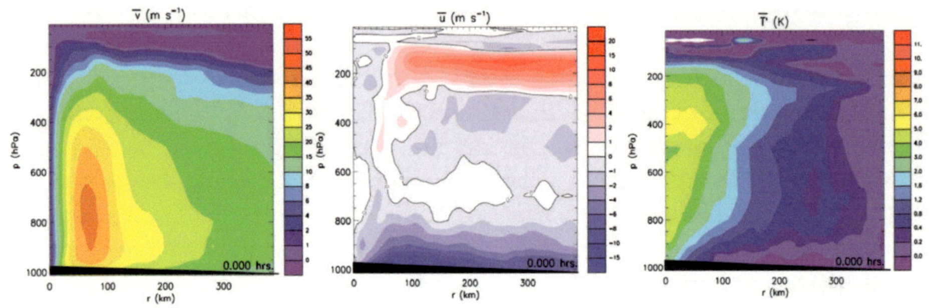

Figure. 2. Azimuthal mean structure of the initial condition of Hurricane Bill (2009) in the Naval Research Laboratory's Coupled Ocean/Atmosphere Mesoscale Prediction System COAMPS®. Panels: a) tangential velocity (m s^{-1}), b) radial velocity (m s^{-1}), and c) perturbation temperature (K). Reproduced from [18].© Copyright 2011 AMS (http://www.ametsoc.org/pubs/crnotice.html).

Using the quasi-balance approximation, where the vorticity is much larger than the divergence, the *f*-plane radial momentum equation can be approximated by

$$\frac{\partial \Phi}{\partial r} = \frac{v^2}{r} + fv, \tag{1}$$

where $\Phi = gz$ is the geopotential, v is the tangential velocity, f is the Coriolis parameter, and r is the radius from the TC center. Outside of deep convective regions, the hydrostatic approximation (in pressure coordinates) is also largely valid,

$$\frac{\partial \Phi}{\partial p} = -\frac{RT}{p}, \tag{2}$$

where p is the pressure, R is the gas constant, and T is the air temperature. Taking $\partial/\partial p$ (1) and $\partial/\partial r$ (2) while eliminating the mixed derivative term, the vortex thermal wind relation is obtained

$$\frac{\partial v}{\partial p}\left(\frac{2v}{r} + f\right) = -\frac{R}{p}\frac{\partial T}{\partial r}. \tag{3}$$

This equation states that a vortex in which v decreases with decreasing p must have warm core, i.e., T must decrease with increasing radius. This is evident in Fig. 2b, where the warm core begins at upper levels, where v is rapidly decreasing.

http://dx.doi.org/10.5772/51177

In the outflow and boundary layers, there exists significant divergent and convergence, respectively, such that the quasi-balance approximation is no longer valid. Therefore an appropriate initialization scheme for TCs should not only capture the primary axisymmetric tangential (azimuthal) circulation, but also the secondary circulation, including the boundary and outflow layers. Additionally, there must be a thermodynamic balance between the boundary layer inflow, rising air in deep and shallow convection, and upper level outflow.

2.2. Asymmetric structure

In order to illustrate some asymmetric features in TCs, Fig. 3 shows two hurricanes: Hurricanes Dolly (2008) and Alex (2010). Hurricane Dolly was very asymmetric in the inner-core region. Note the azimuthal wavenumber-4 pattern in the eyewall radar reflectivity. Hurricane Alex (2010) was also very asymmetric, and had a large spiral rainband emanating from the core, and no visible eye. The point illustrated here is that TCs come in a wide variety of shapes and sizes, and often have prominent asymmetric features. While there is some structure dependence on intensity (i.e., stronger TCs in general are more axisymmetric than weaker TCs), at any initial time a given TC may have very different structure, and the goal of the initialization system is to capture its true state. Remote satellite measurements generally give a decent estimate of the horizontal structure. In fact, microwave data has allowed the ability to "see through" visible and infrared cloud shields, giving improved estimates of the deep convection and precipitation. However, typically there is much less data about the vertical structure. For example, the boundary layer structure or convective and stratiform heating profiles of Alex's rainband would not generally be known. Due to the lack of observations in TCs, in TC initialization systems, aspects of the structure are often specified using estimated information from satellite images.

Figure. 3. Radar and visible satellite imagery depicting asymmetric features in TCs. Hurricane Dolly (2008) (left panel) had asymmetries in the eyewall and rain bands. Hurricane Alex (2010) (right panel) had a large azimuthal wavenumber-1 spiral rain band propagating outward from the vortex center. The left panel is courtesy of the NOAA National Weather Service and the right panel is courtesy of the NOAA/NESDIS in Fort Collins, CO.

3. Direct insertion schemes

As discussed in the previous section, TCs are poorly observed, particularly in the inner-core region. The North Atlantic basin is the only basin that routinely has aircraft reconnaissance missions into storms when they are close to the U.S. southeast coastal regions. The aircraft reconnaissance missions can provide important inner-core structural data using airborne Doppler radar and dropwindsondes, as well as direct or remote measurements of surface wind speed and minimum central pressure. Due to the lack of observations of the inner-core structure of TCs, vortex "bogussing" has been used to improve the representation of the TC in numerical prediction systems. Generally speaking, vortex bogussing is the creation of a TC-like vortex that can be inserted into the initial fields of numerical models [28]. The direct insertion methods take a bogus vortex and insert it directly into the numerical model initial conditions. The bogus vortex can be generated in different ways, which are described below. The main strength of these methods is that the vortex is usually self-consistent. However, some weaknesses exist. First, there can be imbalances that may exist when blending the inserted vortex with the environments in the model analysis. Secondly, for weak TCs and TCs experiencing vertical shear, it is not desirable to insert a vertically stacked vortex into the initial conditions (which is often the case with bogus vortices). Additionally previous studies have shown strong sensitivity to the vertical structure of the bogus vortex, which is often not well observed [46].

After a bogus vortex is created, there needs to be a method to properly insert this vortex into the initial fields of the forecast model. The first guess fields (or the previous model forecast which is valid at the analysis time), usually will already contain a TC-like vortex from the previous forecast. However this vortex may have an incorrect position, intensity, and structure, and therefore it should be removed from model fields. Vortex removal and insertion methods require a number of steps. The common method, discussed by [26] is as follows. First, the total field (e.g., surface pressure) is decomposed into a basic field and disturbance field using filtering. Next, the vortex with specified length scale is removed from the disturbance field. Then, the environmental field is constructed by adding the non-hurricane disturbance with the basic field. Finally, the specified vortex can then simply be added to the environmental field. Schemes of this nature are widely used in operational tropical cyclone prediction models in order to improve the TC representation from the global analysis [27, 34, 50].

3.1. Static vortex insertion

Since TCs are observed to largely be in gradient and hydrostatic balance above the boundary layer [49], one method is to insert a balanced vortex. Routine warning messages are generated by TC warning centers that include estimates of the maximum sustained surface wind, central pressure, and size characteristics (such as the radii of 34 kt winds). Using a function fit to the observed radial wind profile (e.g., a modified Rankine vortex or more sophisticated methods [19, 20]) along with a vertical decay assumption, one can obtain an axisymmetric tangential wind field in the radius-height plane. Following this, the mass field (temperature and pressure) may be obtained by solving the nonlinear balance equation in conjunction with the hydrostatic equation. Then this balanced vortex may be directly inserted into the model initial conditions, as a representation of the actual observed TC vortex. While this method is relatively straightforward, there are a few potential problems: (i) TC vortices are not balanced in the boundary and outflow layers, where strong divergence exists, and (ii) in convectively active regions of the vortex the hydrostatic balance assumption is not valid. It is possible to relax the strict balance assumptions above by building in the boundary layer and outflow structure diagnostically. The addition of boundary and outflow layers should reduce the amount of initial adjustment after insertion.

3.2. Insertion of a dynamically initialized vortex

Instead of specifying a vortex (usually analytically) to represent a TC, another method is to spin-up a TC-like vortex in a numerical model in an environment with no mean flow, and then insert this vortex into the model initial conditions. This method is called a TC dynamic initialization method because the TC vortex is developed from numerical simulation of a nonlinear atmospheric prediction model with full physics that requires prior model integration. The benefits of such a procedure are that the numerical model will generate a more realistic structure for the boundary layer and the outflow layer, and the moisture variables can also be included. The TC dynamic initialization is usually accomplished through Newtonian relaxation. A Newtonian relaxation term is added to the right hand side of a desired prognostic variable (e.g., the tangential velocity or surface pressure) in order to anchor the vortex to the desired structure and/or intensity. The Geophysical Fluid Dynamics Laboratory hurricane prediction model uses an axisymmetric version of its primitive equation to perform the dynamic initialization to a prescribed structure [3, 26, 27]. Recent work has also shown encouraging results with the TC dynamic initialization method using an independent three-dimensional primitive equation model in conjunction with a three-dimensional variational (3DVAR) data assimilation scheme [18, 61]. In Fig. 4, a flow diagram is shown depicting a TC dynamic initialization method applied after three-dimensional variational (3DVAR) data assimilation, where TCs are spun up using Newtonian relaxation to the observed surface pressure. This procedure showed a positive improvement in TC intensity prediction, as average errors in maximum sustained surface wind and minimum central pressure were reduced at all forecast lead times.

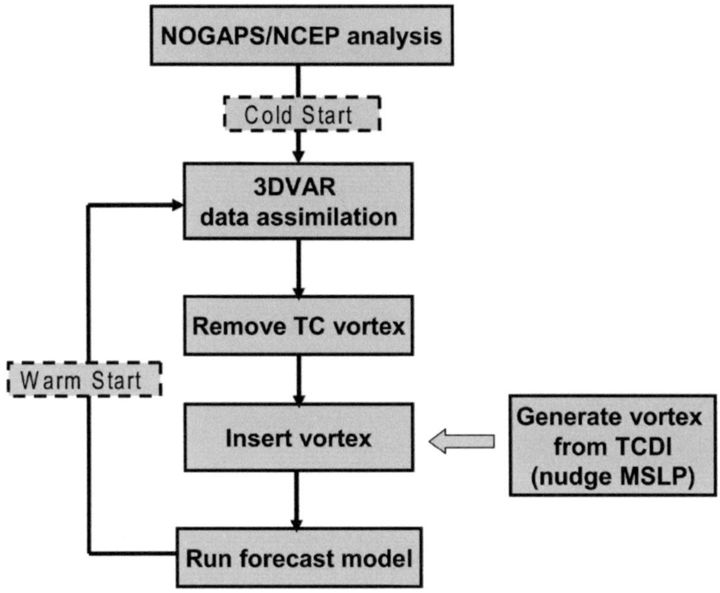

Figure. 4. Application of a TC dynamic initialization scheme to a 3DVAR system, reproduced from [18]. A TC is nudged to observed central mean sea level pressure (MSLP) in a nonlinear full-physics model, and then inserted into the forecast model initial conditions after 3DVAR. © Copyright 2011 AMS (http://www.ametsoc.org/pubs/crnotice.html)

4. Data assimilation systems for TC initialization

The purpose of data assimilation is to produce initial states (analyses) for numerical prediction that maximizes the use of information contained in observations and prior model forecasts to produce the best possible predictions of future states. Most data assimilation methods use observations (e.g., in-situ and remote measurements) to correct short-term model forecasts (the first guess), and therefore the accuracy of the resulting analysis is not just a function of the data assimilation methodology, but the fidelity of the forecast model itself. This analysis is then used as the initial condition for the forecast model. In this section, we discuss the data assimilation strategies that incorporate observational data into the model for proper representation of TCs at the initial time.

In the variational method, a cost function is minimized to produce an analysis that takes into account both the model and observation (including instrument and representativeness) errors. 3DVAR systems (or three-dimensional variational methods) solve this cost function in the three spatial dimensions, while 4DVAR (four-dimensional) systems add the temporal component in a set window. Generally speaking, most atmospheric observations are more applicable to the synoptic scale flow pattern, and often there are few (if any) observations of the inner-core of TCs or other mesoscale or small scale phenomena, aside from infrequent field campaigns. Yet even if these observations exist, it is not trivial to assimilate them while ensuring the proper vortex dynamic and thermodynamic balances.

4.1. 3DVAR systems

The replacement of optimal interpolation (OI) data assimilation scheme by the variational (VAR) method significantly improved the forecast skill of numerical weather prediction systems. The motivation originated from the difficulties associated with the assimilation of satellite data such as TOVS (TIROS-N Operational Vertical Sounders) radiances. It was shown by [31] that the statistical estimation problem could be cast in a variational form (3DVAR) which is a different way of solving the problem than the OI scheme which solves directly. The first implementation of 3DVAR was done at the National Centers for environmental Prediction (NCEP) [36] and later on at the European Center for Medium Range Weather Forecasting (ECMWF) [4]. Other centers like the Canadian Meteorological Centre [13], the Met Office [30], and Naval Research Laboratory [6] also implemented a 3DVAR scheme operationally.

The common method for TC vortex initialization in 3DVAR systems is through the use of adding synthetic observations [15, 17, 29, 55, 65]. Synthetic observations are observations that are created from the estimates of the TC structure and intensity that come from tropical cyclone warning centers (such as the National Hurricane Center in Miami, FL, and the Joint Typhoon Warning Center in Pearl Harbor, HI), and give the best estimate of the storm position, intensity and structure. The synthetic observations are used to enhance the TC representation in the numerical model initial conditions, which generally cannot be adequately captured using the conventional observations. The synthetic observations themselves may be created by sampling a function that matches the observed vortex, and these observations are treated as radiosonde data with assigned proper position information and are included with all other observations and blended with the model first guess using the 3DVAR system. Generally speaking, the observation error is set very low with the TC synthetic observations in the assimilation process, so that the analysis process will largely retain these characteristics of the synthetic observations near the TC. A number of TC synthetic observations are shown for Typhoon Morakot (2009) in Fig. 5, which are ingested into the Naval Research Laboratory's 3DVAR scheme [6], reproduced from [29].

One strength of 3DVAR systems is that synthetic or other TC observations from reconnaissance missions can be assimilated easily into the system. The main problem with using 3DVAR systems for TC initialization is that they generally do not have the proper balance constraints for mesoscale phenomena. Most 3DVAR systems have a geostrophic balance condition to relate the mass and wind fields, which is not valid for tropical cyclones and other strongly rotating mesoscale systems, where there exists a nonlinear balance between the mass and wind fields. The improper balance constraint for TCs in 3DVAR systems can result in rapid adjustment during the first few hours of model integration, causing the model vortex to deviate to a state that is very different from the initially ingested synthetic observations. This discrepancy will most likely be carried throughout the forecast period and can cause a large bias for intensity prediction. It has been recently demonstrated how quickly a 3DVAR system can lose the desired TC characteristics [61]. Additionally, it is very hard to use a 3DVAR data assimilation system to adequately capture the secondary circulation correctly, so as to have consistency between the boundary-layer inflow, vertical motion and heating, and outflow.

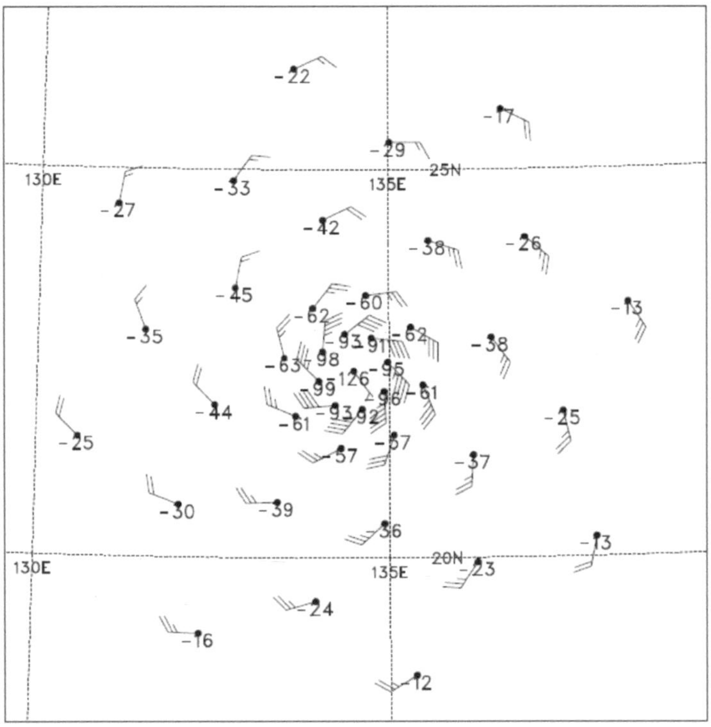

Figure. 5. Depiction of near-surface TC synthetic observations for Typhoon Morakot (2009), reproduced from [29]. The synthetic TC observations are blended with all other observations in the 3DVAR data assimilation.

In addition to the synthetic data, dropwindsonde data from aircraft reconnaissance missions may also be included in variational data assimilation systems. Dropwindsondes measure a quasi-vertical profile of the troposphere from where they are launched. A number of studies have shown a positive impact of assimilating dropwindsonde data on TC track [47, 51]. However there can be significant variability on the impact on a case by case basis.

4.2. 4DVAR systems

The 4DVAR data assimilation system is a generalization of 3DVAR for assimilating observations that are distributed within a specified time window. The goal of 4DVAR is to significantly improve the 3DVAR deficiencies, especially in properly initializing a multi-scale weather system. Compared to 3DVAR, the 4DVAR analyses do not typically show a significant imbalance in the first hours of the forecast. This spin-up process is often associated with the presence of spurious gravity waves that need to be removed by an initialization process (discussed in the next section). A 4DVAR data assimilation system usually requires the development of the tangent linear model and corresponding adjoint system for the forecast model, which are not trivial, in order to iteratively minimize the difference between the first guess fields and the observation. 4DVAR data assimilation systems have been developed for major operation centers for their global prediction system and have led to improvements in forecast skill: ECMWF [40], the Canadian Meterological Centre [14], the U.K. Met Office [41], the Naval Research Laboratory [56], and the Australian Bureau of Meteorology. In some of the 4DVAR systems, synthetic observations are also ingested to improve the TC vortex representation, similar to 3DVAR systems.

An example of an operational TC prediction model that uses a 4DVAR scheme for initialization is ACCESS-TC (Australian Community Climate and Earth System Simulator system for Tropical Cyclones), and a number of other studies have also employed 4DVAR systems for TC initialization [35, 52, 54, 63, 64]. For example, the utility of 4DVAR data assimilation in assimilating irregularly distributed observations in both space and time (such as AMSU-A retrieved temperature and wind fields, as well as the mean sea level pressure (MSLP) information) has been shown by [63]. Using a 72-hour simulation of a land-falling typhoon, they concluded that both the satellite data and the MSLP information could improve the typhoon track forecast, especially for the recurving of the track and landing point. The MM5-4DVAR data assimilation system developed by the Air Force Weather Agency (AFWA) [42] has been employed [62] with a comprehensive satellite products to construct a continuous-coverage, high-resolution TC dataset. Twelve typhoons that occurred over the western Pacific region from May to October 2004 were selected for this reanalysis. The resulting analysis fields show very similar structure of TCs in comparison with satellite observations, demonstrating the capability of 4DVAR in retaining the final structure of the data.

4.3. Ensemble Kalman filter systems

Another four-dimensional data assimilation system, the ensemble Kalman filter (EnKF), has also been adopted for geophysical models [11, 21]. The Kalman filter, is an algorithm which uses a series of measurements observed over time (thus four-dimensional), produces estimates of unknown variables. More formally, the Kalman filter operates recursively on streams of noisy input data to produce a statistically optimal estimate of the underlying system state. The original Kalman Filter assumes that all probability density functions are Gaussian and provides algebraic formulas for the change of the mean and the covariance matrix by the Bayesian update, as well as a formula for advancing the covariance matrix in time provided the system is linear. However, maintaining the covariance matrix is not computationally feasible for high-dimensional systems. For this reason, EnKFs were developed that replace the covariance matrix by the sample covariance computed from the ensemble forecast. The EnKF is now an important data assimilation component of ensemble forecasting. An overview of the work done with the EnKF in the oceanographic and atmospheric sciences can be found in [12].

An intercomparison of an EnKF data assimilation method with the 3D and 4D Variational methods was made using the Weather Research and Forecasting (WRF) model over the contiguous United States during June of 2003 [60]. It is found that 4DVAR has consistently smaller errors than that of 3DVAR for winds and temperature at all forecast lead times except at 60 and 72 h when their forecast errors become comparable in amplitude. The forecast error of the EnKF is comparable to that of the 4DVAR at the 12-36 h lead times, both of which are substantially smaller than that of the 3DVAR, despite the fact that 3DVAR fits the sounding observations much more closely at the analysis time. The advantage of the EnKF becomes even more evident at the 48-72 h lead times.

The EnKF has recently been applied to the TC initialization problem [1, 9, 16, 44, 45, 48, 53, 58, 59]. The EnKF assimilation of inner-core data, such as airborne Doppler radar winds has shown some promising results with improving the vortex structure and intensity forecasts [1, 57]. In Fig. 6, the performance of an EnKF system for predicting TC intensity is shown for a sample of cases in which airborne Doppler radar data was assimilated, reproduced from [57]. As shown in the figure, average intensity errors were reduced by the EnKF assimilation of radar data. [53] used an ensemble Kalman filter (EnKF) to assimilate center position, velocity of storm motion, and surface axisymmetric wind structure in a high-resolution mesoscale model during the 24-h initialization period to develop a dynamically balanced TC vortex without employing any extra bogus schemes. The surface radial wind profile is constructed by fitting the combined information from both the best-track and the dropwindsonde data available from aircraft surveillance observations, such as the Dropwindsonde Observations for Typhoon Surveillance near the Taiwan Region (DOTSTAR). The subsequent numerical integration shows minor adjustments during early periods, indicating that the analysis fields obtained from this method are dynamically balanced. While the EnKF methods are appealing, due to its ensemble nature, it can be significantly more costly (in a computational sense) than the variational methods.

5. Initialization Schemes

While the direct insertion and data assimilation techniques can produce estimates of the observed TC, inevitably imbalances will exist after interpolation and analyses procedures. As discussed earlier, the imbalances will typically be greater for the 3DVAR schemes than 4D schemes. The primary purpose of the initialization schemes is to improve the initial dynamic and thermodynamic balances of the TC, so that spurious gravity waves are filtered from the initial condition [5]. In this section, we discuss three widely used initialization schemes: nonlinear normal mode initialization, digital filters, and dynamic initialization.

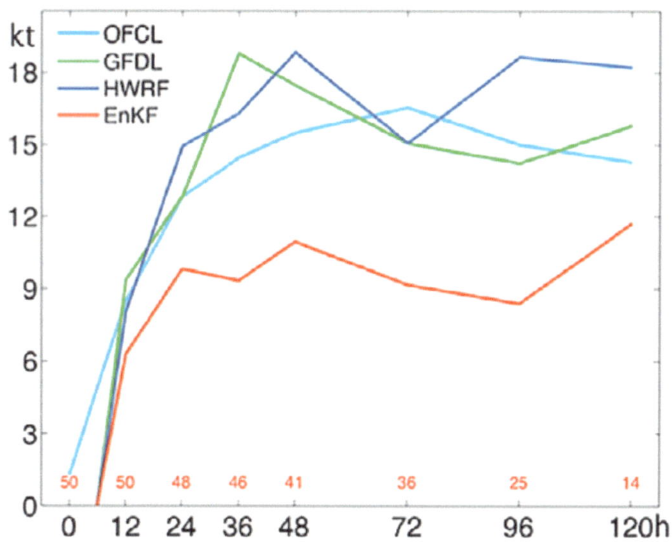

Figure. 6. Mean absolute error (ordinate) in the maximum sustained surface wind versus forecast lead time (abscissa) in a homonegeous sample of cases with airborne Doppler radar data during 2008-2010. As shown the EnKF system which assimilates the radar data had a lower average intensity error than the offical National Hurricane Center forecast (OFCL) and other operational hurricane prediction models (GFDL and HWRF). Figure is courtesy of Fuqing Zhang, reproduced from [57] by permission of American Geophysical Union.

5.1 Nonlinear normal mode initialization

Since an important goal of initialization to provide a balanced initial state from which minimum spurious gravity activity remains [5], methods have been specifically developed to remove such gravity waves from the initial conditions. An early strategy for removal of high frequency oscillations is the nonlinear normal mode method [2, 33, 43]. The eigenvalues of the linearized version of the nonlinear forecast model are the normal modes of the system. For a three-dimensional atmospheric model, these normal modes will encompass higher frequency sound and gravity waves, as well as lower frequency Rossby waves. The idea with the normal mode initialization is to project the analysis vector on to the slower modes in order to reduce gravity waves in the initialization.

5.2 Digital filters

Another method to remove high frequency variability is the digital filter. Similar to the electronic analogue, the digital filter performs a mathematical operation on a time signal to reduce or enhance certain aspects of that signal. For atmospheric applications, this is usually accomplished using a filter that has a cutoff frequency, so that waves of a desired frequency can be removed from the analysis [32]. The benefits of the digital filter is that it is a straightforward way to remove waves of a certain frequency without changing the initial condition significantly [22]. The digital filter can be used in both adiabatic and diabatic modes.

5.3 Dynamic initialization

Dynamic initialization (DI) is a short-term integration of the full model before it actually starts the forecast integration to allow the forecast model to handle the spin-up issue. It usually includes two steps: adiabatic backward integration (i.e., to −6 hour) and diabatic forward integration to the initial time. During adiabatic backward integration, the model physics does not contribute to the tendency of the variables so that this process is quasi-reversible (except the effect of numerical diffusion). In the forward integration (i.e., from −6 hour to the actual initial time at zero hour), the model incurs diabatic process with Newtonian relaxation to some chosen variables so that the initial fields are close to the analysis without introducing small model error during the extra integration time. The idea here is, taking TC prediction as an example, that the 3DVAR procedure produced a reasonably accurate initial state, however, imbalances for TCs with their multiple scales will exist and they should be removed prior to the start of model integration. This process also allows for the build up of the boundary layer and secondary circulation of the TC. The forward DI can be accomplished by relaxation to any or a combination of the model prognostic variables at the analysis time. Of course, much care should be taken in choosing the proper combination. One commonly adopted DI procedure is to relax to the analysis horizontal momentum during the initialization period. DI can also be enhanced by separately relaxing to the nondivergent and divergent wind components, with different relaxation coefficients [7]. This is useful because the nondivergent winds are better captured by the 3DVAR analysis than the divergent winds, and allows for direct way of including relaxation to the heating profiles (which affect the divergent circulation). Various methods have used to incorporate the diabatic effects into the dynamic initialization procedure. These methods include modifying the humidity vertical profiles due to rain rate assimilation, physical initialization, and dynamic nudging to the satellite observed heating profiles [7, 23, 24, 25, 37, 38, 39]. As an example of an operational system, the Australian Bureau of Meteorology used a diabatic dynamic initialization scheme in their earlier tropical cyclone prediction system (TC-LAPS). The diabatic, dynamic initialization was used after a high-resolution objective analysis to improve the mass-wind balance of the vortex while building in the heating asymmetries [8].

6. Conclusions

This chapter reviewed different methods for initializing TCs in numerical prediction systems. The methods range from simpler direct insertion techniques to more advanced dynamic initialization, and from three-dimensional to four-dimensional data assimilation techniques. The strengths and weaknesses of the different schemes were discussed. The direct insertion techniques take either an analytically specified vortex or a dynamically initialized vortex and insert it into the numerical model analysis. These schemes require removal of the TC vortex in the numerical model first guess or analyzed fields, which is often not at the right location or does not match the observations. The direct insertion schemes are appealing because a vortex can be constructed to match the observations, however, there is no guarantee that when inserting this vortex into the analysis that dynamic and thermodynamic balance will exist. In the data assimilation techniques for TC initialization, synthetic observations matching the observed TC structure and intensity are created, and a data assimilation system blends these observations with all other observations to generate the analysis. 3DVAR systems are not as well suited for the TC initialization due to its inability to produce a nonlinear balance between the mass and wind fields. 4DVAR and ensemble Kalman filter schemes show some promising results for TC initialization, in particular, in obtaining a better dynamic and thermodynamic balance, and in the case of the EnKF also providing probabilistic information by running an ensemble. Finally, full domain dynamic initialization (adiabatic and diabatic) techniques were discussed. These schemes are advantageous because they are relatively straightforward to implement, and they are able to produce better dynamic and thermodynamically balanced vortices without the development of the four-dimensional data assimilation.

There are a number of significant challenges that remain for TC initialization. First, most TCs lack of observations needed to construct accurate structure for the storms. Only a handful of TCs in the North Atlantic Ocean basin have routine reconnaissance missions. No matter how advanced the initialization system is, it will always be limited by lack or uncertainty in the observations. Secondly, TCs span multiple scales of motion, ranging from turbulence to deep convective updrafts to vortex scale waves (e.g. vortex Rossby waves), to its interaction with the environments and synoptic scale features. While the synoptic scale is largely responsible for TC track, many of these smaller-scale features are important for intensity. These features are transient and unbalanced, leading to initialization challenges. Third, it is difficult to initialize TCs properly in different environments, such as a TC in shear or with dry air wrapping into its core. Finally, if TC intensity largely depends on deep convective evolution, there are inherent limits to predictability.

In spite of these challenges, much progress has been made of the TC initialization front, and there are promising results from the EnKF, 4DVAR and dynamic initialization schemes. The recent trend in data assimilation is to combine the advantages of 4DVAR and the Kalman filter techniques. Considering the threat that TCs will continue to play, efforts must continue to develop enhanced initialization schemes along with the new technologies for data assimilation to better predict track and intensity.

Acknowledgements

This research is supported by the Chief of Naval Research through the NRL Base Program, PE 0601153N. The authors thank Jim Doyle and Jon Moskaitis for their comments and assistance.

Author details

Eric A. Hendricks* and Melinda S. Peng

*Address all correspondence to: eric.hendricks@nrlmry.navy.mil

Marine Meteorology Division, Naval Research Laboratory, Monterey, CA, USA

References

[1] Aksoy, Altug, Lorsolo, Sylvie, Vukicevic, Tomislava, Sellwood, Kathryn J., Aberson, Sim D., & Zhang, Fuqing. (2012). The HWRF hurricane ensemble data assimilation system (HEDAS) for high-resolution data: The impact of airborne Doppler radar observations in an OSSE. *Mon. Wea. Rev. in press.*

[2] Baer, F., & Tribbia, J. J. (1977). On complete filtering of gravity modes through nonlinear initialization. *Mon. Wea. Rev.*, 105, 1536-1539.

[3] Bender, Morris A., Ross, Rebecca J., Tuleya, Robert E., & Kurihara, Yoshio M. (1993). Improvements in tropical cyclone track and intensity forecasts using the GFDL initialization system. *Mon. Wea. Rev.*, 121, 2046-2061.

[4] Courtier, P., Andersson, E., Heckley, W., Pailleux, J., Vasiljevic, D., Hamrud, M., Hollingsworth, A., Rabier, F., & Fisher, M. (1998). The ECMWF implementation of three-dimensional variational assimilation (3D-Var). Part 1: Formulation. *Quart. J. Roy. Meteor. Soc.*, 124, 1783-1807.

[5] Daley, Roger. (1991). Atmospheric data analysis. Cambridge University Press.

[6] Daley, Roger, & Barker, Edward. (2001). NAVDAS: Formulation and diagnostics. *Mon. Wea. Rev.*, 129, 869-883.

[7] Davidson, Noel E., & Puri, Kamal. (1992). Tropical prediction using dynamical nudging, satellite-defined convective heat sources, and a cyclone bogus. *Mon. Wea. Rev.*, 120, 2329-2341.

[8] Davidson, Noel E., & Weber, Harry C. (2000). The BMRC high-resolution tropical cyclone prediction system: TC-LAPS. *Mon. Wea. Rev.*, 128, 1245-1265.

[9] Dong, Jili, & Xue, Ming. Assimilation of radial velocity and reflectivity data from coastal WSR-88D radars using ensemble Kalman filter for the analysis and forecast of landfalling Hurricane Ike (2008). *Quart. J. Roy. Met. Soc. in press.*

[10] Dvorak, Vernon F. (1975). Tropical cyclone intensity analysis and forecasting from satellite imagery. *Mon. Wea. Rev.*, 103, 420-430.

[11] Evensen, Geir. (1994). Sequential data assimilation with a nonlinear quasi-geostrophic model using Monte Carlo methods to forecast error statistics. *J. Geophys. Res.*, 99, 143-162.

[12] Evensen, Geir. (2003). The ensemble Kalman filter: theoretical formulation and practical implementation. *Ocean Dynamics*, 53, 343-367.

[13] Gauthier, Pierre, Charette, C., Fillion, L., Koclas, P., & Laroche, S. (1999). Implementation of a 3D variational data assimilation system at the Canadian Meteorological Centre. Part I: The global analysis. *Atmosphere-Oceans*, 37, 103-156.

[14] Gauthier, Pierre, Tanguay, Monique, Laroche, Stephane, Pellerin, Simon, & Morneau, Josee. (2007). Extension of 3DVAR to 4DVAR: Implementation of 4DVAR at the Meteorological Service of Canada. *Mon. Wea. Rev.*, 135, 233-2354.

[15] Goerss, James S., & Jeffries, Richard A. (1994). Assimilation of synthetic tropical cyclone observations into the Navy Operational Global Atmospheric Prediction System. *Wea. Forecasting*, 9, 557-576.

[16] Hamill, Thomas M., Whitaker, Jeffrey S., Fiorino, Michael, & Benjamin, Stanley G. (2011). Global ensemble predictions of 2009's tropical cyclones initialized with an ensemble Kalman filter. *Mon. Wea. Rev.*, 139, 668-688.

[17] Heming, J. T., Chan, J. C. L., & Radford, A. M. (1995). A new scheme for the initialisation of tropical cyclones in the UK Meteorological Office global model. *Meteor. Appl.*, page DOI: 10.1002/met.5060020211.

[18] Hendricks, Eric A., Peng, Melinda S., Li, Tim, & Xuyang, Ge. (2011). Performance of a dynamic initialization scheme in the Coupled Ocean-Atmosphere Mesoscale Prediction System for Tropical Cyclones (COAMPS-TC). *Wea. Forecasting*, 26, 650-663.

[19] Holland, Greg J. (1980). An analytic model of the wind and pressure profiles in hurricanes. *Mon. Wea. Rev.*, 108, 1212-1218.

[20] Holland, Greg J. (2008). A revised hurricane pressure-wind model. *Mon. Wea. Rev.*, 136, 3432-3445.

[21] Houtemaker, P. L., & Mitchell, H. L. (1998). Data assimilation using an ensemble Kalman filter technique. *Mon. Wea. Rev.*, 126, 796-811.

[22] Huang, Xiang-Yu, & Lynch, Peter. (1993). Diabatic digital-filtering initialization: Application to the HIRLAM model. *Mon. Wea. Rev.*, 121, 589-603.

[23] Krishnamurti, T. N., Bedi, H. S., Heckley, William, & Ingles, Kevin. (1988). Reduction in spinup time for evaporation and precipitation in a spectral model. *Mon. Wea. Rev.*, 116, 907-920.

[24] Krishnamurti, T. N., Correa-Torres, Ricardo, Rohaly, Greg, Oosterhof, Darlene, & Surgi, Naomi. (1997). Physical initialization and hurricane ensemble forecasts. *Wea. Forecasting*, 12, 503-514.

[25] Krishnamurti, T. N., Han, Wei, Jha, Bhaskar, & Bedi, H.S. (1998). Numerical prediction of Hurricane Opal. *Mon. Wea. Rev.*, 126, 1347-1363.

[26] Kurihara, Yoshio M., Bender, Morris A., & Ross, Rebecca J. (1993). An initialization scheme of hurricane models by vortex specification. *Mon. Wea. Rev.*, 121, 2030-2045.

[27] Kurihara, Yoshio M., Bender, Morris A., Tuleya, Robert E., & Ross, Rebecca J. (1995). Improvements in the GFDL Hurricane Prediction System. *Mon. Wea. Rev.*, 123, 2791-2801.

[28] Leslie, Lance M., & Holland, G. J. (1995). On the bogussing of tropical cyclones in numerical models: A comparison of vortex profiles. *Meteorol. Atmos. Phys.*, 56, 101-110.

[29] Liou, C. S., & Sashegyi, Keith D. (2012). On the initialization of tropical cyclones with a three-dimensional variational analysis. *Natural Hazards*, 63, 1375-1391.

[30] Lorenc, A. C., Ballard, S. P., Bell, R. S., Ingleby, N. B., Andrews, P. L. F., Barker, D. M., Bray, J. R., Clayton, A. M., Dalby, T., Li, D., Payne, T. J., & Saunders, F. W. (2000). The Met. Office global three-dimensional variational data assimilation scheme. *Quart. J. Roy. Meteor. Soc.*, 126, 2991-3012.

[31] Lorenz, A. (1986). Analysis methods for numerical weather prediction. *Quart. J. Roy. Meteor. Soc.*, 112, 1177-1194.

[32] Lynch, Peter, & Huang, Xiang-Yu. (1992). Initialization of the HIRLAM model using a digital filter. *Mon. Wea. Rev.*, 120, 1019-1034.

[33] Machenhauer, B. (1977). On the dynamics of gravity oscillations in a shallow water model, with applications to normal mode initialisation. *Beitr. Phys. Atmos.*, 50, 253-271.

[34] Mathur, Makut B. (1991). The National Meteorological Center''s quasi-Lagrangian model for hurricane prediction. *Mon. Wea. Rev.*, 119, 1419-1447.

[35] Park, Kyungjeen, & Zou, X. (2004). Toward developing an objective 4DVAR BDA scheme for hurricane initialization based on TPC observered parameters. *Mon. Wea. Rev.*, 132, 2054-2069.

[36] Parrish, David F., & Derber, John C. (1992). The National Meteorological Center''s spectral statistical-interpolation analysis system. *Mon. Wea. Rev.*, 120, 1747-1763.

[37] Peng, Melinda S., & Chang, Simon W. (1996). Impacts of SSM/I retrieved rainfall rates on numerical prediction of a tropical cyclone. *Mon. Wea. Rev.*, 124, 1181-1198.

[38] Peng, Melinda S., Jeng, B. F., & Chang, C. P. (1993). Forecast of typhoon motion in the vicinity of Taiwan during 1989-90 using a dynamical model. *Wea. Forecasting*, 8, 309-325.

[39] Puri, K., & Davidson, N. E. (1992). The use of infrared satellite cloud imagery data as proxy data for moisture and diabatic heating in data assimilation. *Mon. Wea. Rev.*, 120, 2329-2341.

[40] Rabier, F., Jarvinen, H., Klinker, E., Mahfouf, J.-F., & Simmons, A. (2000). The ECMWF operational implementation of four-dimensional variational assimilation. I: Experimental results with simplified physics. *Quart. J. Roy. Meteor. Soc.*, 126, 1143-1170.

[41] Rawlins, F., Ballard, S. P., Bovis, K. J., Clayton, A. M., Li, D., Inverarity, G.W., Lorenc, A.C., & Payne, T. J. (2006). The Met Office global four-dimensional variational data assimilation scheme. *Quart. J. Roy. Meteor. Soc.*, 133, 347-362.

[42] Ruggiero, F. H., Michalakes, J., Nehrkorn, T., Modica, G. D., & Zou, X. (2006). Development of a new distributed-memory MM5 adjoint. *J. Atmos. Ocean Tech.*, 23, 424-436.

[43] Temperton, C. (1988). Implicit normal mode initialization. *Mon. Wea. Rev.*, 116, 1013-1031.

[44] Torn, Ryan D. (2010). Performance of a mesoscale ensemble Kalman filter (EnKF) during the NOAA high-resolution hurricane test. *Mon. Wea. Rev.*, 138, 4375-4392.

[45] Torn, Ryan D., & Hakim, Greg J. (2009). Ensemble data assimilation applied to RAINEX observations of Hurricane Katrina (2005). *Mon. Wea. Rev.*, 137, 2817-2829.

[46] Wang, Yuqing. (1998). On the bogusing of tropical cyclones in numerical models: The influence of vertical tilt. *Meteorol. Atmos. Phys.*, 65, 153-170.

[47] Weissmann, Martin, Harnisch, Florian, Chun-Chieh, Wu, Lin, Po-Hsiung, Ohta, Yoichiro, Yamashita, Koji, Kim, Yeon-Hee, Jeon, Eun-Hee, Nakazawa, Tetsuo, & Aberson, Sim. (2011). The influence of assimilating dropsonde data on typhoon track and midlatitude forecasts. *Mon. Wea. Rev.*, 139, 908-920.

[48] Weng, Yonghui, & Zhang, Fuqing. (2012). Assimilating airborne Doppler radar observations with an ensemble Kalman filter for convection-permitting hurricane initialization and prediction: Katrina (2005). *Mon. Wea. Rev.*, 140, 841-859.

[49] Willoughby, Hugh E. (1990). Gradient balance in tropical cyclones. *J. Atmos. Sci.*, 47, 265-274.

[50] Winterbottom, Henry R., & Chassignet, Eric P. (2011). A vortex isolation and removal algorithm for numerical weather prediction model tropical cyclone applications. *J. Adv. Model. Earth. Sys.*, 3(M11003), 8.

[51] Chun-Chieh, Wu, Chou, Kun-Hsuan, Lin, Po-Hsiung, Aberson, Sim D., Peng, Melinda S., & Nakazawa, Tetsuo. (2007). The impact of dropwindsonde data on typhoon track forecasts in DOTSTAR. *Wea. Forecasting*, 22, 1157-1176.

[52] Chun-Chieh, Wu, Chou, Kun-Hsuan, Wang, Yuqing, & Kuo, Ying-Hwa. (2006). Tropical cyclone initialization and prediction based on four-dimensional variational data assimilation. *J. Atmos. Sci.*, 63, 2383-2395.

[53] Chun-Chieh, Wu, Lien, Guo-Yuan, Chen, Jan-Huey, & Zhang, Fuqing. (2007). Assimilation of tropical cyclone track and structure based on the ensemble Kalman filter (EnKF). *J. Atmos. Sci.*, 67, 3806-3822.

[54] Zhao Xia, Pu, & Braun, Scott A. (2001). Evaluation of bogus vortex techniques with four-dimensional variational data assimilation. *Mon. Wea. Rev.*, 129, 2023-2039.

[55] Xiao, Qingnong, Kuo, Ying-Hwa, Zhang, Ying, Barker, Dale M., & Won, Duk-Jin. (2006). A tropical cyclone bogus data assimilation scheme in the MM5 3D-Var system and numerical experiments with Typhoon Rusa (2002) near landfall. *J. Meteor. Soc. Japan*, 84, 671-689.

[56] Liang, Xu, Rosmond, Tom, & Daley, Roger. (2005). Development of NAVDAS-AR: Formulation and initial tests of the linear problem. *Tellus*, 57A, 546-559.

[57] Zhang, Fuqing, Weng, Yonghui, Gamache, John F., & Marks, Frank D. (2011). Performance of convection-permitting hurricane initialization and prediction during 2008-2010 with ensemble data assimilation of inner-core airborne Doppler radar observations. *Geophys. Res. Lett.*, 38, L15810.

[58] Zhang, Fuqing, Weng, Yonghui, Kuo, Ying-Hwa, Whitaker, Jeffrey S., & Xie, Baoguo. (2010). Predicting Typhoon Morakot"s catastrophic rainfall with a convection-permitting mesoscale ensemble system. *Wea. Forecasting*, 25, 1861-1825.

[59] Zhang, Fuqing, Weng, Yonghui, Sippel, Jason A., Meng, Zhiyong, & Bishop, Craig H. (2009). Cloud-resolving hurricane initialization and prediction through assimilation of Doppler radar observations with an ensemble Kalman filter. *Mon. Wea. Rev.*, 137, 2105-2125.

[60] Zhang, Meng, Zhang, Fuqing, Huang, Xiang-Yu, & Zhang, Xin. (2011). Intercomparison of an ensemble Kalman filter with three- and four-dimensional variational data assimilation methods in a limited-area model over the month of June 2003. *Mon. Wea. Rev.*, 139, 566-572.

[61] Zhang, Shengjun, Li, Tim, Xuyang, Ge, Peng, Melinda S., & Pan, Ning. (2012). A 3DVar-based dynamical initialization scheme for tropical cyclone predictions. *Wea. Forecasting*, 27, 473-483.

[62] Zhang, X., Li, T., Weng, F., Wu, C. C., & Xu, L. (2007). Reanalysis of western Pacific typhoons in 2004 with multi-satellite observations. *Meteorol. Atmos. Phys.*, 97, 3-18.

[63] Zhao, Y., Wang, B., Ji, Z., Liang, X., Deng, G., & Zhang, X. (2005). Improved track forecasting of a typhoon reaching landfall from four-dimensional variational data assimilation of AMSU-A retrieved data. *J. Geophys. Res.*, 110(D14101).

[64] Zhao, Ying, Wang, Bin, & Liu, Juanjuan. (2012). A DRP-4DVar data assimilation scheme for typhoon initialization using sea level pressure data. *Mon. Wea. Rev.*, 140, 1191-1203.

[65] Zou, X., & Xiao, Q. (2000). Studies on the initialization and simulation of a mature hurricane using a variational bogus data assimilation scheme. *J. Atmos. Sci.*, 57, 836-860.

Chapter 2

Elaboration of Technologies for the Diagnosis of Tropical Hurricanes Beginning in Oceans with Remote Sensing Methods

A. G. Grankov, S. V. Marechek, A. A. Milshin,
E. P. Novichikhin, S. P. Golovachev,
N. K. Shelobanova and A. M. Shutko

Additional information is available at the end of the chapter

http://dx.doi.org/10.5772/53863

1. Introduction

Satellite passive microwave (MCW) radiometric methods are an important tool for determining the oceanographic and meteorological parameters that affect the energy exchange in the ocean-atmosphere system (SOA), such as sea surface temperature, wind speed, the total amount of water vapor in the atmosphere, integral water vapor content of the clouds, precipitation intensity, and also especially important to study the characteristics of the cyclonic areas of the ocean are the vertical turbulent fluxes of heat, moisture and momentum. These satellite measurements can also give indirect information about the factors important from the point of formation of tropical storms processes in the ocean and on its bottom, outside of direct line of sight remote means. Their use in this case "allows researcher to look into the ocean column on the surface of which as a kind of screen are projected the various images of deep-water processes" [1].

The long-term goal of this research is the creation of methods and technologies for diagnosing the origin of tropical hurricanes (THs) in the areas of the ocean, which are regular sources, the origin of hurricanes on basis of the data of passive MCW microwave radiometric measurements from satellite, ship, buoy measurements and results of mathematical modeling of the behavior of the parameters of the ocean-atmosphere system (SOA) at different stages: the stage that precedes the appearance of TH; the appearance TH; the stage of SOA relaxation after TH appearance.

The important theoretical purpose of the work is the search for effects and regularities, which can explain the reasons and circumstances under which THs appearance is inevitable.

2. Behavior of parameters of the atmosphere immediately before the appearance of hurricanes

In this section are presented some results of the study of the SOA reaction on passing the powerful Hurricane Katrina in August 2005 in the Florida Straits in the area of the buoy station SMKF1 (Sombrero Key) as well as the results of a behavior of the system in the period of time preceding the beginning and development of the Hurricane Humberto in the Gulf of Mexico in September 2007 at the point of the buoy station 42019. For these time periods an analysis of the following synoptic variations of atmospheric and oceanic characteristics were conducted: These include air temperature, humidity, pressure and wind speed in the near-surface 10-th meter layer in area of the stations SMKF1 and 42019, vertical turbulent fluxes of sensible and latent heat at the sea-water boundary calculated with the measurement data the stations SMKF1 and 42019 and integral (total) water vapor content and enthalpy of the atmosphere calculated by integration of the air humidity and temperature within the height range 10-10000 meters.

The source of information on the earth-based data is the American center - National Data Buoy Center NOAA (NDBC); the data of regular measurements from the microwave radiometers, SSM/I (Special Sensor Microwave/Imager) of the meteorological satellite F15 DMSP and AMSR-E (Advanced Microwave Scanning Radiometer) of the satellite EOS Aqua were used as the source of satellite data. The technical characteristics of these radiometers are given in references [2] and [3], respectively.

Spacious network of meteorological of the NOAA stations, in particular, the stations situated in the Gulf of Mexico and the equatorial zone of the Pacific Ocean provide exclusively measurements of the parameters of the ocean surface and near-surface atmosphere. Meteorological means of observation from these stations are not able to give information on the vertical distribution of temperature and humidity in the atmosphere. This problem can be solved by means of use of measurement data of the multichannel MCW radiometer SSMIS (Special Sensor Microwave Imager/Sounder) from satellites DMSP F16 and F17 [4]. In addition to this function of the scanner, this device is able to determine the atmosphere temperature and humidity at various heights. However, a periodicity of remote sensing these atmospheric characteristics (once per day) is not enough for studying such fast processes as tropical hurricane formation, with noticeably varying characteristics during several hours.

The method of combining the data of the buoy measurements of the atmospheric near-surface layer and the ocean surface parameters with data obtained from satellite MCW measurements has been developed, which provide information on the air temperature and humidity, not only in the near-surface atmosphere but also in overlying atmospheric layers. This technique allows the determination of values of the atmosphere temperature and humidity at its various horizons (the property of satellite MCW radiometric measurements) and hourly (the property of buoy meteorological measurements).

3. Dynamics of meteorological parameters measured from the stations SMKF1 and 42019

3.1. Station SMKF1 (TH Katrina)

The station SMKF1 from the NDBC data arsenal is used as the reference point in the Florida Straits (24.38° N, 81.07° W) when analyzing an influence of the Hurricane Katrina on the atmospheric parameters. The nearest distance between a trajectory of Katrina and this station was ~120 km in the noon of 26 August 2005, by this moment the hurricane has passed about 800 km from the place of its formation near the Bahamas.

The NDBC data in an area of the station SMKF1 between 21 and 31 August 2005 was analyzed. It can be observed here that significant contrasts of the near-surface air parameters with respect to their undisturbed (background) values are appearing before the coming of Hurricane Katrina and after it moving off: the variations of the air temperature, humidity and pressure are about -6°C, -15 mb and -13 mb, respectively.

Fig. 1 illustrates variations of the air temperature t_a and pressure P in the atmosphere near-surface layer between the 21st and 31st August 2005 recorded by sensors of the SMKF1 station as well as computed values of the near-surface air humidity (water vapor pressure) e. These results were obtained using the data from previous studies of the relationship of the parameter e with the difference of water and air temperatures in various zones of the world ocean derived in [5]: the NOAA buoy stations are not includes direct measurements of the air humidity. Fig. 1 presents the smoothed results of station measurements of the parameters t_a, P and calculated estimates of the parameter e. A smoothing is compiled with the standard means of the computer program ORIGIN Adjacent Averaging with the 3-hour interval of averaging of the hourly samples. Initial data level from the SMKF1 sensors was 240 hourly samples for each of the parameters t_a, e and P, characterizing the stage proceeding an appearance of the Hurricane Katrina in an area of the station SMKF1 (21-24 August), the stage of its passing this area (25-29 August) and the stage of the SOA relaxation (30-31 August). These results suggest the need to apply an idea of the explosive effects in the atmosphere during the THs activity.

Results of the linear regression analysis show close interrelations between variations of the near-surface air temperature and humidity, the coefficient of correlation of the parameters t_a and e is 0.94. On the basis of the data of buoy meteorological measurements and using the technique cited in [6] computed values of internal energy (enthalpy) of the near-surface atmosphere in the period from 21 to 31 August 2005 were carried out. When passing the point SMKF1, Hurricane Katrina collects the heat energy from the atmosphere near-surface layer, according to this estimate it is reducing roughly to 32500 J/m^2 in this period.

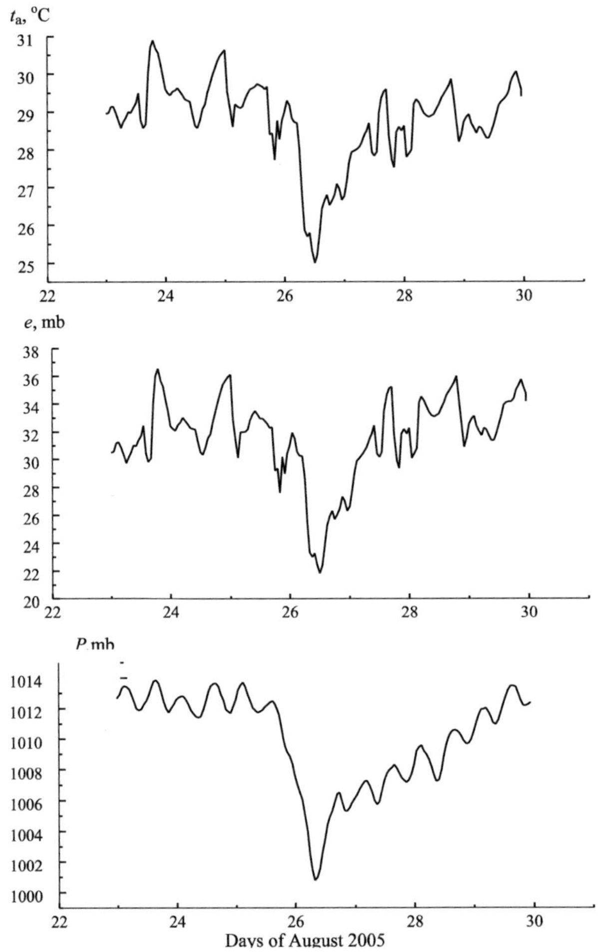

Figure 1. Variations of the near-surface temperature t_a, humidity e, pressure P in the area of location of the station SMKF1 in the Florida Strait during passing the TH Katrina in August 2005.

3.2. Station 42019 (TH Humberto)

Hurricane Humberto was born in the middle of September 2007 in the Gulf of Mexico, it was not as intensive as Hurricane Katrina, but it is important in these studies as its source area coincided with the location of the buoy station 42019 situated at coordinates 27.91° N, 95.35° W. This peculiarity allows the monitoring of parameters of the atmospheric near-surface layer (as well as parameters of overlying layers when using data of simultaneous MCW radiometric measurements) over various stages of forming the hurricane. According to the data measurements from the station 42019 this point is characterized by a strong changeability of parameters of the atmospheric near-surface layer in the period of forming Hurricane Humberto: variations of the air temperature, humidity, pressure, and wind speed amounted to 3°C, 8 mb, 5 mb and 7 m/s, respectively (see Fig. 2).

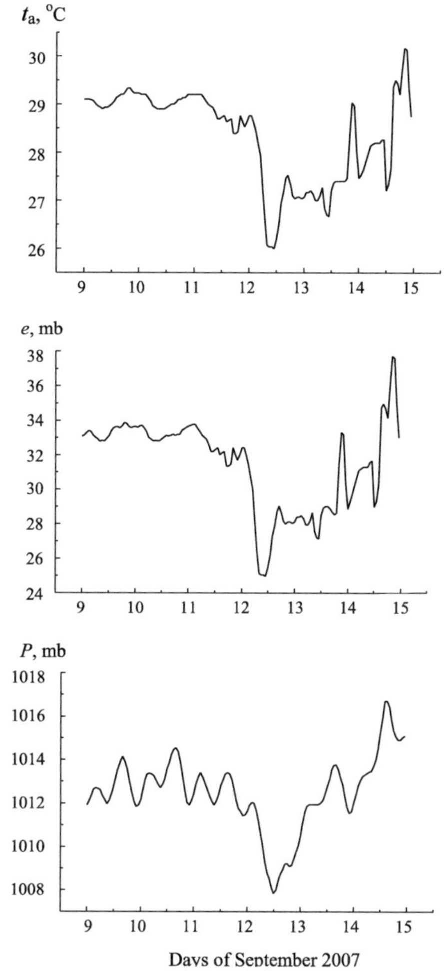

Figure 2. Variations of the near-surface temperature t_a, humidity e, pressure P from the measurement data of the station 42019 in the Gulf of Mexico during the starting the TH Humberto in September 2007.

Variations of the near-surface air humidity in the period from 9th to 14th September practically repeat variations of the near-surface air temperature: the coefficient of their correlation is 0.97. The near-surface air pressure is sharply declining at the stage of this hurricane development (12 September). The atmospheric near-surface layer enthalpy was computed between 9th and 14th September 2007 in the area of location of the station 42019: it follows from the results of computation that the enthalpy has been reduced by 12500 J/m^2 during the development of Hurricane Humberto. An analysis of variations of the ocean surface temperature during the passing the Hurricane Katrina passed the station SMKF1 (22-31 August 2005) and during the period of formation and development of the Hurricane Humberto (8-16 September 2007) has been fulfilled - these results are shown at the Fig. 3. To emphasize the character of behavior of the ocean surface temperature, the data of buoy measurements are approximated with the standard means of the computer technique ORIGIN (Sigmoidal), which produces the stick-slip motion of original dependencies. Figure 3 demonstrates the "jump" of the ocean surface temperature values in area of the station SMKF1 caused by passing the Hurricane Katrina is in a few times more in comparison with this phenomena observed during beginning the Hurricane Humberto.

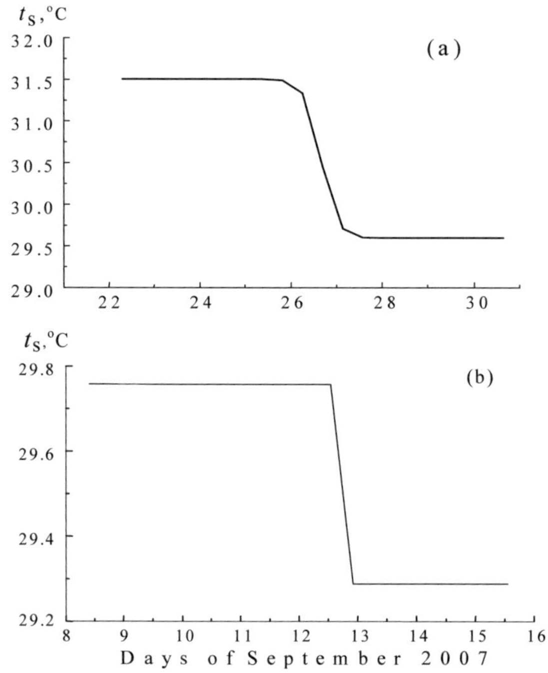

Figure 3. Character of changes of the ocean surface temperature t_s: (a) an area of the station SMKF1 during passing the TH Katrina: (b) an area of the station 42019 in the period of forming the TH Humberto.

4. Dynamics of the surface heat and moisture fluxes

Resting upon the data of buoy measurements of the ocean surface temperature, the near-surface air humidity estimates and wind speed we computed the values of sensible q_h and latent q_e heat at the air-sea boundary using the well-known in dynamic meteorology formulas of the Global Aerodynamic Method) - so called Bulk Formulas were justified in [7]. Due to this approach the values q_h and q_e are characterized with the following relationships:

$$q_h = c_p \rho c_t (t_s - t_a) V; \quad (1)$$

$$q_e = L \rho (0.622 / P) c_e (e - e_o) V, \quad (2)$$

i.e. they are become apparent through following parameters of the SOA - the air temperature t_a, pressure P, humidity e and wind speed V in the near-surface atmosphere, as well as through the ocean surface temperature t_s and proper for this the maximal value of the air humidity e_0. As the constant of proportionality in these relations are served the numbers of Schmidt c_t (heat exchange), Dalton c_e (moisture exchange), the specific heat of evaporation (L), the specific air heat under constant pressure (c_p), and its density (ϱ). Below some results of computing the heat fluxes with reference to the stations SMKF1 and 42019 based on the buoy measurements in these areas of the Gulf of Mexico are presented.

4.1. Station SMKF1

Figure 4 shows some results of computation of the heat fluxes (with 3-hour averaging). One can observe an influence of passing the Hurricane Katrina through the station SMKF1 seen in the form of appreciable reducing of the heat fluxes, about 20 W/m^2 (from 30 to 10 W/m^2) for the fluxes of sensible heat, and about 150 W/m^2 (from 350 to 200 W/m^2) for the fluxes of latent heat. This result demonstrates an effect of smoothing the heat contrasts between the ocean surface and near-surface atmosphere due to the effects of the passing of the hurricane.

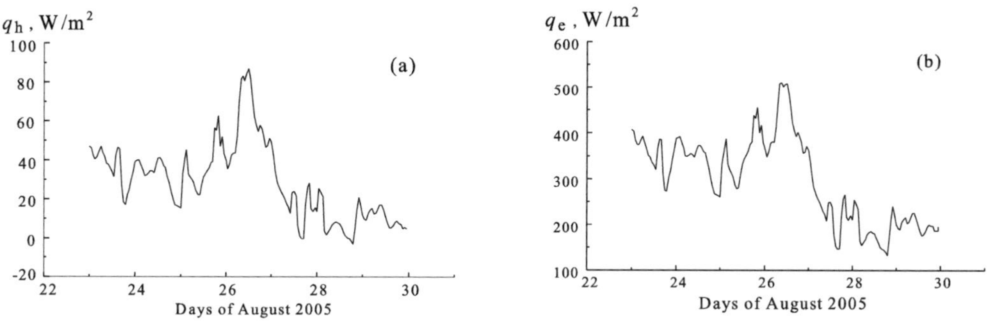

Figure 4. Variations of sensible (a) and latent (b) heat fluxes at the ocean surface in an area of the station SMKF1 location in the period of passing the TH Katrina in August 2005.

The moment of passing of the hurricane over the station SMKF1 (noon of 26 August) is accompanied by a positive increase of the parameters q_h and q_e, which amount to 80 and 500 W/m^2, respectively.

4.2. Station 42019 (Hurricane Humberto)

Figure 5 shows results of computing the fluxes of sensible and latent heat and impulse (with the 3-hour smoothing); one can observe here a sharp maximum peak of the values q_h and q_e simultaneously, which falls at the noon of 12 September 2007 that coincides with the data of ground observations of the Hurricane Humberto development.

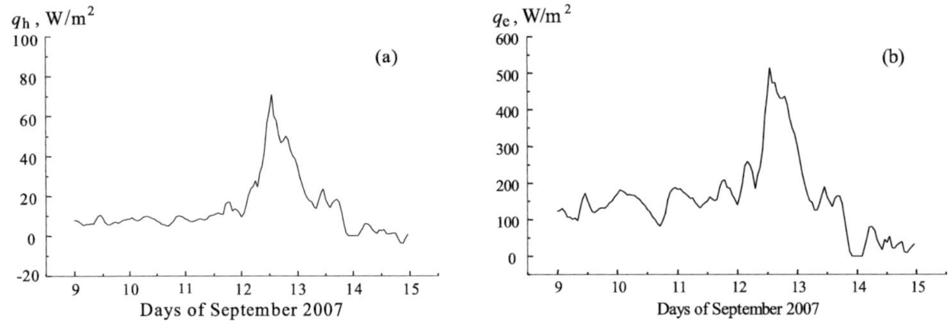

Figure 5. Variations of sensible (a), latent (b) heat and impulse (c) fluxes at the ocean surface in area of the station 42019 location in period of beginning the Hurricane Humberto in September 2007.

Average values of the heat and moisture fluxes at the stage preceding TH Humberto appearance (9-12 September) amount to 5 W/m^2, 150 W/m^2 and 0.05 N/m^2, respectively, and their maximal values at the stage of its development in the noon of 12 September reach to 75 W/m^2, 530 W/m^2 and 0.2 N/m^2. Notably, that maximal value of the total (sensible+latent) heat fluxes in area of the station 42019 (~ 600 W/m^2) is close to the estimate cited by Golytsin for tropical latitudes [8]. Also, this value is comparable with the total heat fluxes values in the Newfoundland energy active zone of the North Atlantic, which is subjected regularly to influence of powerful mid-latitude cyclones, which in compliance with the data of experiments NEWFOUEX-88 and ATLANTEX-90 reached to values of 800 W/m^2 in March 1988 and April 1990 [9].

Figure 6 demonstrates variations of the heat and moisture fluxes in the period 17-20 September at the stage of relaxation of the SOA parameters in area of the station 42019 after the development of Hurricane Humberto and it's leaving this area.

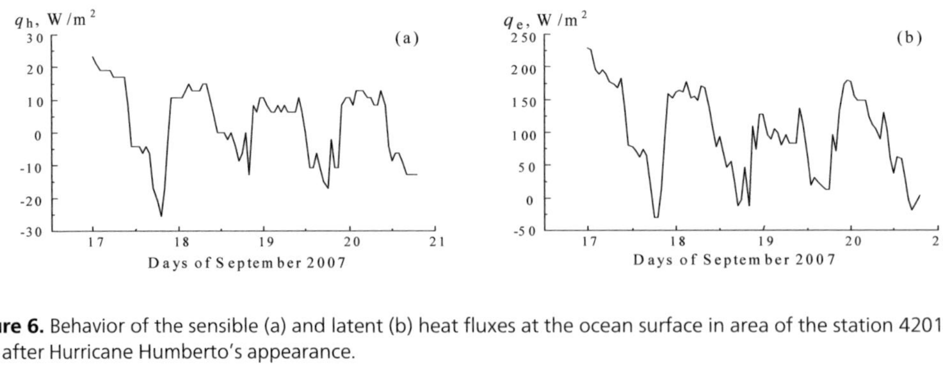

Figure 6. Behavior of the sensible (a) and latent (b) heat fluxes at the ocean surface in area of the station 42019 location after Hurricane Humberto's appearance.

It is seen from the illustration that average values of the parameters q_h and q_e are a few times under their limit values observed at noon on the 12th September. One interesting peculiarity manifests itself - the oscillatory character of variations in the heat and moisture fluxes in this time with the oscillation period closed to 24 hours, i.e. to the diurnal cycle. In addition, the sensible heat fluxes are alternating, that is the processes of heat transfer from the ocean surface to the atmosphere are alternating with the processes of heat transfer from the atmosphere to the ocean surface; this phenomenon was not observed in the period between 9th and 12th September preceding the appearance of Hurricane Humberto (see Fig. 2). This effect is similar to the effect of excitation of oscillations in high-Q resonant systems as the ringing circuits in radio-engineering described in [10] for example.

5. Dynamics of integral water vapor content and enthalpy of the atmosphere

5.1. Technique of determination of the temperature and humidity of the atmospheric upper layers

The sought dependences $t_a(h)$ and и $\varrho(h)$ are found in the form of exponential functions $t_a(h) = t_a(0) \exp(-\kappa_t h)$; $\varrho(h) = \varrho(0) \exp(-\kappa_\varrho h)$ providing a minimal root-mean-square error (discrepancy) between measured by the MCW radiometers SSM/I и AMSR-E values of the SOA brightness temperatures and their simulated (model) estimates. With the dependences $t_a(h)$ and $\varrho(h)$ the linear and integral absorption of radiowaves as well as the brightness temperatures of the SOA natural MCW radiation in various atmospheric layers for all satellite MCW radiometric channels are computed using the-known plane-layer model of natural microwave radiation of the system [11, 12].

As the radiometers SSM/I and AMSR-E are the multi-channel systems, their measurement data seems to be sufficient for determination of the coefficients κ_t and κ_ϱ required for retrieving the dependencies $t_a(h)$ and $\varrho(h)$ over the ocean. The value of discrepancy between simulated and measured estimates of the SOA brightness temperature is computed both with ascending as descending satellite orbits falling into cells 0.25° x 0.25° centralized about the stations SMKF1 and 42019 for the following spectral and polarization channels of the radiometers SSM/I and AMSR-E: a) 37 GHz (0.81 cm), 19 GHz (1.58 cm), vertical and horizontal polarizations; 22.235 GHz (1.35 cm), vertical polarization (radiometer SSM/I and b) 36.5 GHz (0.82 cm), 18.7 GHz (1.6 cm), 23.8 GHz (1.26 cm), vertical and horizontal polarization (radiometer AMSR-E).

This developed technique allows the computation of approximately values of the temperature and humidity of the atmosphere at various horizons for estimating its integral characteristics such as the integral water vapor content and enthalpy (heat content), for example. It seems that mainly the atmosphere integral characteristics will be informative in an analysis of the SOA dynamics in zones of activity of the tropical hurricanes in spite of the fact that the real profiles of the air temperature and humidity can be appreciably different from the exponential ones.

Resting upon the buoy data on the air humidity in the near-surface layer and the computed estimates of this parameter in overlying atmosphere layers the integral water vapor content of the atmosphere (IVA) Q in the layer 10-10000 m was computed. Comparing the results of computing the parameter Q with its satellite estimates derived with the radiometer SSM/I in area of the station SMKF1 and the radiometer AMSR-E in area of the station 42019 was made. Besides, the calculation estimates of the atmosphere enthalpy for various its layers were obtained in areas of activity of the Hurricanes Katrina and Humberto.

5.2. Dynamics of IVA in area of the station SMKF1

Figure 7 compares the estimates of the IVA variations in area of the station SMKF1 during the period 1-30 August 2005 computed by a layer-wise integrating of the air humidity at various heights with the satellite estimates of the parameter Q derived with the measurement data from the radiometer SSM/I using the known technique [13]. It can be observed here the appreciable variations of the parameter Q, which are coincide with the time of passing the Hurricane Humberto through the station SMKF1, when the maximum of the near-surface heat fluxes was observed in the noon 25 August.

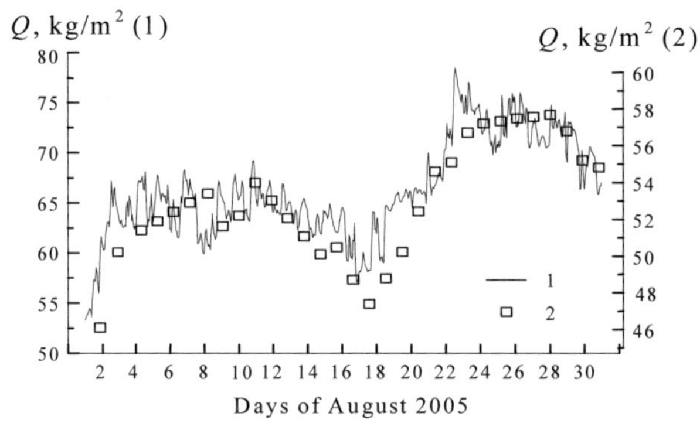

Figure 7. Comparing the estimates of the atmosphere integral water vapor content Q in area of the station SMKF1: 1 - data of computing for the layer 10:10000 m; 2 - data of measurements of the radiometer SSM/I.

A difference between the absolute values of the computed and satellite estimates of the parameter Q can be explained by the fact that when modeling the SOA brightness temperature did not allow for its increase caused by the cloudiness and precipitation, which are registered by the satellite radiometer SSM/I. Irregularity of remote sensing an area of the SMKF1 station and availability of noticeable gaps in the satellite measurements is an important consideration also. Though, one can mark a good compliance of relative changes (variations) of the both estimates; this is essentially for validation of the developed technique of determination of the air temperature and humidity and especially, of their changeability at various horizons of the atmosphere under arising (passing) the tropical hurricanes.

5.3. Dynamics of IVA in area of the station 42019

The IVA estimates was derived in the area of station 42019 in the period 6-15 September including the stages of the formation of Hurricane Humberto and compared the satellite estimates of the parameter Q with the data of measurements from the radiometer AMSR-E (see Fig. 8).

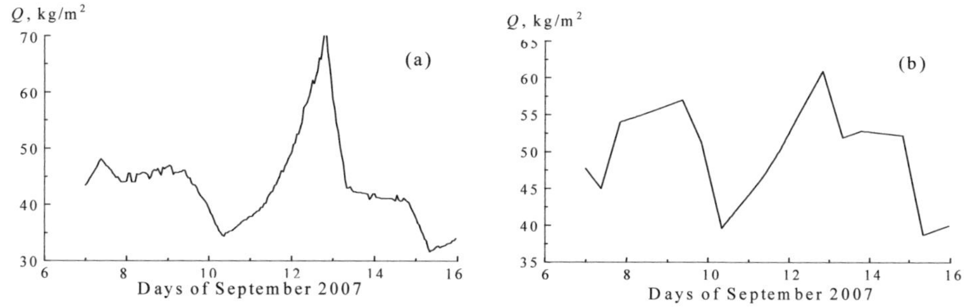

Figure 8. Estimates of the atmosphere integral water vapor content Q in area of the station 42019: (a) - data of computation for the atmosphere layer 10:10000 m; (b) - data of measurements from the radiometer AMSR-E.

It is seen from Fig. 8, that the formation of Hurricane Humberto is accompanied by increases of the value of Q (in the noon of 12 September); one can observe that maximum of Q (in the noon 12 September) happens together with maximums of fluxes of sensible, latent heat. The peak of values of Q on 12th September precedes the fragment (7-9 September) with increased water vapor content of the atmosphere.

Figure 9 compares the satellite and simulated estimates of the parameter Q after Hurricane Humberto left the area of station 42019 for the time interval 13-20 September. It follows from the illustration that a reducing IVA value occurred in this period, which has an oscillatory characteristic, similar to the case of the near-surface fluxes of sensible and latent heat.

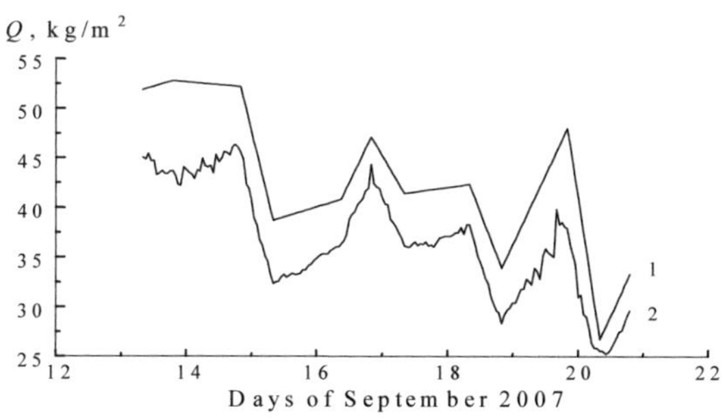

Figure 9. Compare of satellite (1) and simulated (2) estimates of the atmosphere integral water vapor content Q after leaving by the Hurricane Humberto an area of the station 42019.

5.4. Dynamics of the atmosphere enthalpy in areas of the stations SMKF1 and 42019

Variations of the enthalpy (heat content) were computed of various atmospheric layers in areas of the stations SMKF1 and 42019 in periods of activity of the Hurricanes Katrina and

Humberto. Figure 10 demonstrates some results of computation of the enthalpy J for the layer 10–10000 m.

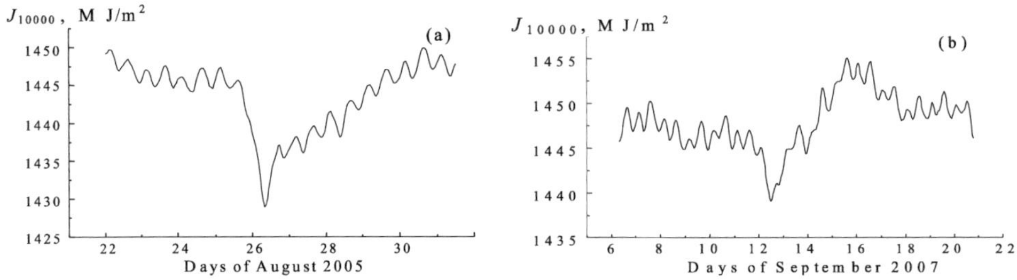

Figure 10. Variations of the atmospheric enthalpy J_{10000} during passing Hurricane Katrina the station SMKF1 (a) and the formation of Hurricane Humberto in area of the station 42019 (b).

This illustration shows that during passing Katrina the station SMKF1 and arising Hurricane Humberto in area of the station 42019 we can observe a sharp reduction of the J_{10000} value. This reduction is of the resonant type and accompanied by a strong increase in the vertical turbulent heat and moisture fluxes at the ocean-atmosphere boundary. Thus, the hurricanes cause a reduction of the atmosphere in these areas occurs, just like this effect occurs for the mid-latitude cyclones in the Newfoundland energy active zone of the North Atlantic in the experiment ATLANTEX-90 as shown in [9].

6. Effect of accumulation of the atmoshere water vapor in the storm situations

The long-term goal is to study some peculiarities of behavior of the atmosphere heat and microwave radiation characteristics in waters observed by contact and distant means for situations preceded to appearance of the storms.

The research areas are: an area of the station SMKF1 (Sombrero Key) in the Florida Straight in August 2005; a shore of the Black Sea (Gelendzik, territory of the South branch of the P.P. Shirshov Institute of Oceanology Russian Academy of Sciences) in September and October 2010 before the appearance of an intensive storm in 1 October).

Using meteorological and satellite microwave radiometric data comparison of the behavior of near-surface air humidity, fluxes of sensible and latent heat, total water vapor content of the atmosphere, as well as the microwave radiation characteristics of the water surface-atmosphere system were carried out in the periods before the appearance of Hurricane Katrina in area of the station SMKF1 and the development of sea storm in the area of the Gelendzik test site This effect is similar to the effect of a heat accumulation in the atmosphere water vapor observed in the Florida Straights in August 2005 when a clear effect of the monotonous strengthening of the integral water vapor content in the atmosphere was observed in the area of station SMKF1 before the atmospheric perturbations caused by appearance the Hurricane Katrina. In both cases the effect of accumulation of the water vapor in the atmosphere was observed.

6.1. The Black Sea experiment: general description and results

Research on the atmosphere parameters were carried out in a near-shore zone of the Black Sea. The measurements were conducted from 11th September until 10th October 2010 in area of the Blue Bay, which is located four kilometers west of Gelendzik Bay. Some data have been derived including for cloudless atmosphere and a calm sea situation and for the situation preceding a storm appearance. Devices of the measuring complex were placed on the pier end (Fig. 11). The pier has a length about 200 m; the depth of water is 7 m.

Besides traditional meteorological sensors a MCW radiometer mounted at the scanning platform was used. The radiometer operated at the wavelength 1.35 cm, i.e. in the line of attenuation of the atmospheric water vapor, which helps estimate a total water vapor content of the atmosphere. The scanning of a hemisphere of space was realized in five fixed directions on an azimuth through 30 degrees. Scanning in an angle of elevation was conducted by slow change of an angle from zenith to a nadir and inverse.

Figure 11. Measuring complex (left) and scanning platform (right).Ю

Figure 12 presents measured scans of the atmosphere brightness temperatures at the wavelength 1.35 cm as the functions of the azimuth angle and the angle of elevation during the period from 23 to 30 September preceding to the storm appearance (the storm has passed the test site on 1 October). It is seen from Fig. 12 the increase of minimal and maximal values of the atmosphere brightness temperature measured under positions of the scanning platform 0 and 90 degrees respectively.

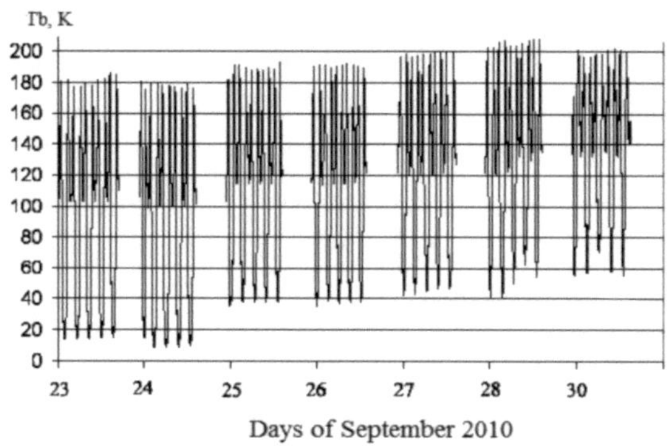

Figure 12. Brightness temperature T^b for the azimuth scans of the atmosphere at the wavelength 1.35 cm before the storm approach into the Blue Bay.

6.2. Dynamics of heat fluxes

Results of meteorological measurements at the pier show the effect of increasing the near-surface air temperature and humidity before the storm came into the Blue Bay area. In this period an accumulation of heat in the near-surface air water vapor occurred; its specific heat content was increased by 61 KJ/m^2. So a sizeable increment of the air enthalpy in comparison with the storm situation in the Florida Straight in August 2005 (when Hurricane Katrina approached station SMKF1) for example is caused by essentially more variations of the temperature and humidity of the near-surface air. The vertical heat and latent fluxes were estimated with the bulk-formulas with use of the direct measurements of the air and water temperature as well as the near-surface wind speed; its results are presented in Fig. 13.

It is seen from Fig.13 that as the storm approaches there is a decrease of the sensible heat flux from 40 to 100 W/m^2 and latent heat flux from 225 to 10 W/m^2 occurred. Before an approach of the sea storm, the water surface gave up its heat to the near-surface air, but about the day before the storm's appearance in Blue Bay a characteristic of the heat interchanges in the water/air interface was changed.

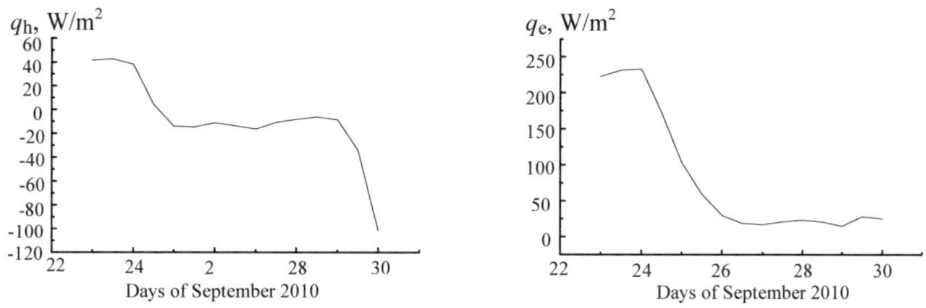

Figure 13. Variations of sensible q_h and latent q_e heat fluxes caused by the storm activity in the Blue Bay.

6.3. Comparison of sea and satellite measurements

A comparison of the variations of the near-surface air temperature and humidity measured directly from the pier by the meteorological sensors with those obtained from the EOS Aqua radiometer AMSR-E at the wavelength 1.26 cm (23.8 GHz) in the sea area bordered by Blue Bay from the Black Sea side was carried out. A comparison of satellite measurement data proper for the Blue Bay will be uncorrected, as its sizes are considerably less in comparison with the spatial resolution of the AMSR-E channel 1.26 cm. Figure 14 show results of comparison of the near-surface air temperature and humidity (a), the brightness temperature measured by the AMSR-E radiometer at the wavelength 1.26 cm (b); the brightness temperature measured from the pier by the radiometer operating at the wavelength 1.35 cm (c). The sizes of selected area are confirmed to the sizes the satellite "spot", which linear sizes for the AMSR-E radiometer 23.8 GHz are of 40x60 km with the center coordinated of 44°W, 38°E.
4

Fig.14 illustrates a similarity of reactions of the SOA characteristics such as the near-surface air temperature and humidity measures from the pier as well as the SOA brightness temperature in the resonant region of the wavelength of attenuation of the radiowaves in the atmospheric water vapor, with any methods of the MCW passive radiometric observation means independently (from the pier or from the space). The effect of "pumping" the near-surface air temperature and humidity before the approach of Hurricane Katrina in the area of the SMKF1 station was observed, also during a progress of the Hurricane Katrina in the Florida Straights (station SMKF1) in August 2005 (Fig. 15). The estimates of the parameter Q were obtained from the measurements of the radiometer AMSR-E at the wavelength 1.26 cm.

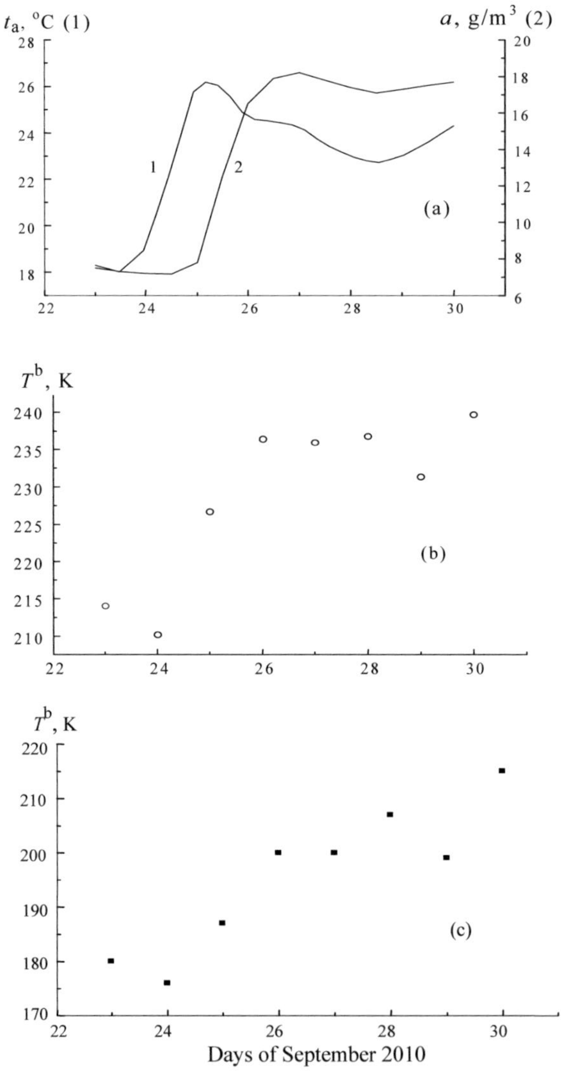

Figure 14. Variations of heat and thermal microwave characteristics of the atmosphere before approach of the sea storm into Blue Bay: (a) - temperature (1) and humidity (2) of the near-surface air; (b) - AMSR-E brightness

temperature at the wavelength 1.26 cm; (c) - brightness temperature measured from the Gelendzik pier at the wavelength 1.35 cm.

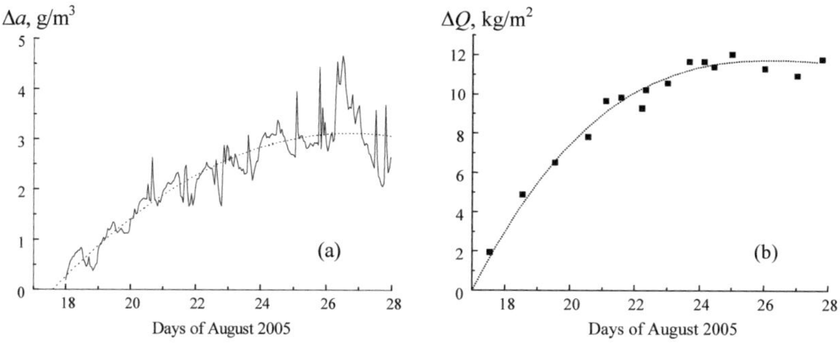

Figure 15. Variations of the absolute humidity of the near surface air Δa (a) and the total water vapor of the atmosphere ΔQ (b) before the approach of Hurricane Katrina to the SMKF1 station. Dot lines - approximation of original results by polynomial of the 2-th order.

7. Conclusions

The dynamics of different characteristics of the atmosphere in periods of activity of Hurricanes Katrina (August 2005) and Humberto (September 2007) was analyzed. This study is based on coupling the data of the buoy meteorological measurements with the data of simultaneous measurements from the DMSP SSM/I radiometer (in area of the NOAA station SMKF1) and the EOS-Aqua AMSR-E radiometer (in area of the station 42019). The technique of analysis of the atmosphere's integral characteristics such as its total water vapor content and enthalpy was developed. It enables determination of variations of the atmosphere temperature and humidity at its various horizons during the passing of Hurricane Katrina passed the station SMKF1 and the formation of Hurricane Humberto in the area of the station 42019. It is shown that in both cases the effect of taking off heat energy by hurricanes from the atmosphere and the ocean surface takes place. This effect results in strong disturbances of the temperature, humidity and pressure in the near-surface atmosphere and is accompanied by a sharp decrease of the atmosphere enthalpy and considerable increase of the vertical turbulent heat and moisture fluxes at the ocean surface.

The following features of the results of an analysis of the SOA parameters dynamics during the process of the formation of Hurricane Humberto can be noted, firstly the oscillating nature of behavior of the sensible and latent heat fluxes as well as the atmosphere integral water vapor after leaving by the formation location of the hurricane, i.e. at the stage of the SOA relaxation, secondly the availability of anomalies in behavior of the atmosphere integral water vapor the 4-5 days before an appearance of Hurricane Humberto. This leaves open to further research the question of what becomes of these anomalies as the hurricane passes. This research is required in order to further examine with observational data the processes of formation of other tropical hurricanes in various oceanic areas in various years and seasons taking into account the effect of the atmospheric horizontal circulation.

A good agreement between simulated and satellite estimates of the variations of the atmosphere water vapor content in areas of the stations SMKF1 and 42019 indicates that the adopted exponential model of vertical distribution of the air temperature and humidity provides useful information about dynamics of this integral characteristic of the atmosphere state, which is important for studying the influence of transient and developing tropical cyclones on the state of the system "ocean-atmosphere".

Results of the researches confirm a role of the atmospheric integral water vapor content, which is ease accessible in studying with modern satellite microwave radiometric means at the stages proceeding to development of the storm situations in the ocean and the sea, this is an important and required condition for studying the problems of the tropical cyclone genesis [14].

Acknowledgements

These investigations were conducted in a frame of the project on elaboration of technologies for the diagnosis of tropical hurricanes beginning in oceans. The project was maintained by the International Scientific and Technological Center (ISTC) in 2008-2011, grant no. 3827. We are grateful to Dr. Vladimir Krapivin, (V.A. Kotelnikov Institute of Radioengineering and Electronics RAS) and Dr E. Sharkov (Institute of Space Reseaches) for collaboration and useful discussions. We are also grateful to Dr. A. Lugovskoi for the assistance in our work.

Author details

A. G. Grankov, S. V. Marechek, A. A. Milshin, E. P. Novichikhin, S. P. Golovachev,
N. K. Shelobanova and A. M. Shutko

*Address all correspondence to: agrankov@inbox.ru

Kotelnikov Institute of Radioengineering and Electronics Russian Academy of Sciences, Moscow, Russia

References

[1] Fedorov KN, Ginsburg AI. The Subsurface Layer of the Ocean, Leningrad, SSSR: Gidrometeoizdat; 1988 (in Russian).

[2] Hollinger PH, Peirce JL, Poe GA. SSM/I Instrument Evaluation. IEEE Transactions Geoscience Remote Sensing 1990 12 781-790.

[3] Kawanishi T, Sezai T. Imaoka K et al. The Advanced Microwave Scanning Radiometer for the Earth Observing System (AMSR-E), NASDA's Contribution to the EOS for Global Energy and Water Cycle Studies. IEEE Transactions Geoscience Remote Sensing 2003 48 173-183.

[4] Kunkee D, Boucher D, G, Poe G, Swadley S. Evaluation of the Defence Meteorological Satellite Program (DMSP) Special Microwave Imager Sounder (SSMIS). In: proceedings of the IEEE International Symposium Geoscience and Remote Sensing Symposium (IGARSS2006) 7-12 July, Denver, Kolorado, 2006.

[5] Snopkov VG. On correlation between the atmosphere water vapor and the near surface humidity seasonal variations of the water vapor content over the Atlantic. Meteorol Gidrol 1977 12 38-42 (in Russian).

[6] Xrgian AX. Physics of the Atmosphere, vol.1. Leningrad, USSR: Gidrometeoizdat; 1978 (in Russian).

[7] Lappo SS, Gulev SK, Rozhdestvenskii AE. Large-scale Heat Interaction in the Ocean-Atmosphere System and Energy-Active Zones in the World Ocean. Leningrad, USSR: Gidrometeoizdat; 1990 (in Russian).

[8] Golytsyn GS. Polar Lows and Tropical Hurricanes: Their Energy and Sizes and a Quantitative Criterion for Their Generation. Izvestija, Atmosp. Oceanic Phys 2008 44(5) 579-590 (in Russian).

[9] Grankov AG, Milshin AA. Microwave Radiation of the Ocean-Atmosphere: Boundary Heat and Dynamic Interaction. Berlin, Germany: Springer-Verlag; 2010.

[10] Kharkevich AA. Bases of the Radioengineering (3-th edition). Moscow, Russia: Fizmatlit; 2007 (in Russian).

[11] Basharinov AE, Gurvich AS, Egorov ST. Radio Emission of the Planet Earth. Moscow, USSR: Fizmatlit (Nauka); 1974 (in Russian).

[12] Armand NA, Polyakov VM. Radio Propagation and Remote Sensing of the Environment. CRC Press LLC: Roca Raton; 2005.

[13] Alishouse JC, Snyder SA, Vongsatorn J, Ferrado RR. Determination of Oceanic Total Precipitate Water from the SSM/I. Journal Geophysical Researches 1990 28(5) 811-816.

[14] Sharkov AJ, Shramkov JN, Pokrovskaya IV. The Integral Water Vapor in Tropical Zone as the Necessary Condition for Atmospheric Catastrophic Genesis. Rem. Sens. Earth from Space. 2012, 2 73-82 (in Russian).

Chapter 3

Assessment of a Parametric Hurricane Surface Wind Model for Tropical Cyclones in the Gulf of Mexico

Kelin Hu, Qin Chen and Patrick Fitzpatrick

Additional information is available at the end of the chapter

http://dx.doi.org/10.5772/51288

1. Introduction

Tropical cyclones, which generate storm surges, wind waves, and flooding at landfall, are a major threat to human life and property in coastal regions throughout the world. The United States, the northern Gulf of Mexico, and in particular the Louisiana Gulf coast, are very susceptible to the impacts of frequent tropical storms and hurricanes due to its tropical/subtropical location and unique bathymetric, geometric, and landscape features. Severe coastal flooding, enormous property damage, and loss of life are ubiquitously associated with tropical cyclone landfalls, and this devastation was no more evident than during Hurricanes Katrina and Rita in 2005, and Gustav and Ike in 2008. Over 1800 people lost their lives and several major coastal populations were crippled for months after the hurricanes passed. It is critically important to make timely and accurate forecasts of hurricane winds, surge, and waves. The prediction of hurricane surface winds is of specific importance because it directly forces storm surge and wave models, and controls their forecast accuracy.

In the past several decades, numerous wind models and products have been developed to hindcast and forecast hurricane surface wind fields. Methodologies include H*Wind kinematic analysis winds [1], steady-state slab planetary boundary (PBL) models [2-3], interactive objective kinematic analysis system which combine steady-state slab PBL models with observations such as IOKA [4], and weather model output (operational models as well as mesoscale research models such as the Weather Research and Forecasting Model, known as WRF). After landfall, an exponentially based filling model for central pressure is sometimes applied to the PBL schemes [5].

However, many applications utilize analytical parametric formulations which represent radial profiles of tropical cyclone winds. The schemes are "parametric" in that the radial wind variation depends on just a few parameters, such as the maximum winds, the radius of maximum wind, and central pressure. The relative simplicity, near-zero computational cost, and flexible grid resolution favors the use of parametric winds for: assessing hurricane wind return periods and risk modeling for insurance underwriting; engineering applications; hindcasting tropical cyclone studies; and forcing of wave and surge models [6-7]. The equations exploit the basic structure of tropical cyclones in which pressure decreases exponentially towards the center then levels off in the eye, while the winds increase exponentially toward the center, then decrease to calm inside the eyewall. Radial wind profiles have evolved from the Rankine combined vortex formulation (where solid-body rotation is assumed inside the eyewall, then tangential winds decrease by a radial scaling parameter) to one in which a rectangular hyperbola approximation to radial pressure variation is used [8]. This basic concept has resulted in many wind profile relationships (e.g., [9-10]).

However, the most popular scheme is based on the 1980 Holland wind profile [11]. Since tropical cyclone winds contain significant radial profile variations, Holland modified the Schloemer equation [8] to represent a spectrum of pressure-varying rectangular hyperbolas:

$$p(r) = p_c + (p_n - p_c)e^{-\left(\frac{R_m}{r}\right)^B} \tag{1}$$

which includes an additional scaling parameter *B*:

$$B = \frac{V_m^2 \rho_a e}{p_n - p_c} \tag{2}$$

and a wind profile given by:

$$V(r) = \left[\frac{B}{\rho_a}\left(\frac{R_m}{r}\right)^B (p_n - p_c)e^{-(R_m/r)^B} + \gamma\left(\frac{rf}{2}\right)^2\right]^x - \gamma\left(\frac{rf}{2}\right) \tag{3}$$

where p is the pressure at radius r, p_c is the central pressure, p_n is the ambient pressure, R_m is the radius of maximum wind V_m, V is the wind at r, ρ_a is the air density, f is the Coriolis parameter ($f = 2\Omega sin\phi$), Ω is the rotational frequency of the earth, ϕ is the latitude, and e is the base of the natural logarithm. B is the wind shape parameter with values typically between 1 and 2.5; for a given intensity, larger B values concentrate more of the pressure drop near R_m and the wind profile becomes more "peaked." Holland [11] assumed cyclostrophic balance where $x = 0.5$ and $\gamma = 0$. However, Hu et al. [12] retained the Coriolis terms ($\gamma = 1$) and showed that excluding the Coriolis effect in the parameter B, that is, using Eq. (2) instead of Eq. (4) to determine B, can lead to 20% errors of V_m for weak but large tropical cyclones, defaulting to gradient wind balance ($V_g = V$).

While popular, the Holland formulation [11] is known to have problems. It cannot represent double eyewalls. Another issue is the inability to accurately represent the eyewall and outer-core winds simultaneously; it can match the outer wind profile accurately but fails to capture the rapid decrease of wind just outside the eyewall (and often underestimates the true V_m), and conversely may represent the eyewall winds well but decrease too rapidly far away from the center [13]. Willoughby et al. [14] proposed an alternative piecewise continuous wind profiles, while others have suggested alternate expressions for B [15-17]. Recently, Holland et al. [6] proposed a cyclostrophic expression ($\gamma = 0$) in which x varies linearly with radius outside the eyewall, and can also be tuned to match observations or bimodal wind profiles.

However, the biggest deficiency with [11] is the 2D nature of the equation, implying that the vortex is symmetric. Actual tropical cyclone wind fields are rarely symmetric, especially for landfalling situations. Xie et al. [18] improved the Holland model by considering asymmetry. Mattocks and Forbes [19] developed an asymmetric wind model based on Xie et al.'s approach [18] and employed it in the storm surge model ADCIRC. In their models, R_m is replaced by a directionally varying $R_m(h)$, where h is the azimuthal angle around the center of the storm. The National Hurricane Center (NHC) forecast advisories and the Automated Tropical Cyclone Forecasting (ATCF) product provide hurricane track and surface (i.e., 10 m) wind forecasts, in which the storm structure is depicted by the radii of specified wind speeds (34, 50, 64, and 100 knots) in four quadrants. The four-quadrant information is used to compute $R_m(h)$ at any azimuth through a fitting procedure. Notice that the gradient wind expression ($x = 0.5$; $\gamma = 1$) in Eq. (3) excludes the translational velocity generated by the moving hurricane. This is inconsistent with the NHC forecast wind data that include both the vortex-related and translational wind speeds. Therefore, the translational portion of the wind speed will have to be removed from the forecast (or observed values) before applying Eq. (3). Moreover, normally for each quadrant, only the largest available specified wind velocity and its radius are used in existing asymmetric parametric wind models. Owing to the complexity of actual hurricanes, it is important and necessary to make full use of all available forecast wind data to improve the accuracy of a parametric hurricane wind model.

A method to blend the Holland scheme [11] with asymmetries was developed by [12], and is described in the next section. This paper assesses Hu et al.'s [12] parametric hurricane surface wind model using field observations of several historical tropical cyclones in the Gulf of Mexico. In-situ measurements of wind speed and wind direction at 12 offshore buoys in the Gulf of Mexico and the hurricane wind hindcasts from the NHC are utilized as the reference to evaluate the predictive skills of the parametric model. Both major hurricanes and large tropical storms are selected. The resultant hurricane wind fields are merged with the available background wind fields to cover the entire Gulf of Mexico and part of the Atlantic basin, which are used to drive basin-scale storm surge and wind wave models. Statistical tools are used to quantify the accuracy of the hindcasted wind fields. A fully coupled storm surge and wind wave modeling system is then employed to demonstrate the importance of the accuracy of the parametric wind model for surge and wave hindcasts/forecasts in comparison with field observations.

2. Model description

Hu et al. [12] improved a parametric hurricane wind model based on the asymmetric Holland-type vortex models. The model creates a two-dimensional surface wind field based on the

NHC forecast (or observed) hurricane wind and track data. Three improvements have been made to retain consistency between the input parameters and the model output and to better resolve the asymmetric structure of the hurricane. Please refer to [12] for details.

First, in determination of the shape parameter B, the Coriolis effect is included, and the range restriction suggested by Holland [11] of 1.0-2.5 is removed. The parameter B is expressed as follows,

$$B = \frac{\left(V_{gm}^2 + V_{gm} R_m f\right)\rho_a e}{p_n - p_c} \tag{4}$$

If the Coriolis effect is neglected ($f = 0$), Eq. (4) returns to Eq. (2), the commonly used form of the parameter B.

Second, the effect of the translational velocity of a hurricane is excluded from the input of specified wind speeds before applying the Holland-type vortex to avoid exaggeration of the wind asymmetry. The translational velocity is added back in at the very end of the procedure.

Third, a new method has been introduced to develop a weighted composite wind field that makes full use of all wind parameters, not just the largest available specified wind speed and its 4-quadrant radii. For each set of specified wind speeds (30, 50 or 64 knots) and their radii, a gradient wind field can be calculated. Obviously, near the radii of one specified wind speed in four quadrants, the wind field based on the same specified wind speed would be much more accurate than other wind fields and should be assigned the most weight in the combined wind field, while simultaneously, the weighting coefficient for other wind fields should approach zero.

3. Recent tropical cyclone cases in the Gulf of Mexico

Ten historical tropical cyclone cases were carried out to test the performance of this wind model. They are Hurricanes Isidore and Lili in 2002, Hurricane Ivan in 2004, Hurricanes Dennis, Emily, Katrina, Rita and Wilma in 2005, and Hurricanes Gustav and Ike in 2008. The hurricane tracks and observation buoys are shown in Figure 1 [20]. The background wind is taken from the 6-hourly NCEP/NCAR Reanalysis data. For 2002 hurricanes, only the 34-knot radii at four quadrants were provided in their best track data (50- and 64-knot radii were unavailable). Due to a lack of other radii information, the wind model was not able to do the weighted composition for Hurricanes Isidore and Lili before merging with the background winds. After a brief introduction of each tropical cyclone case, comparisons of modeled and measured wind speeds/wind directions at available buoys, as well as scatter plots, are provided. Two sets of correlation coefficient (r^2) and root mean square error (RMSE), one for all data and one for observed winds stronger than 10 ms^{-1}, quantify the model performance. The latter set focuses mostly on inner-core winds.

http://dx.doi.org/10.5772/51288

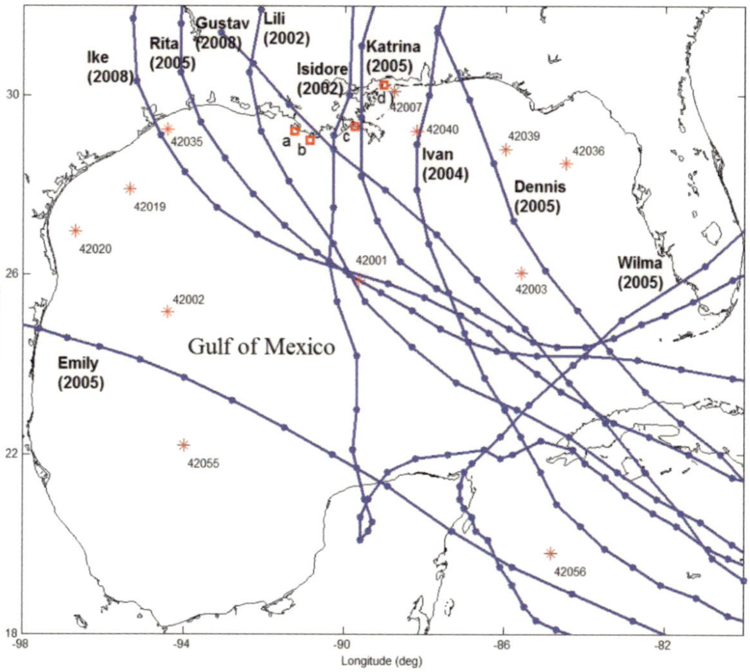

Figure 1. Tracks of ten Gulf of Mexico tropical cyclones, and buoy stations (red stars), and coastal gages (red squares, deployed for Hurricane Gustav) used for validation.

3.1. Hurricane Isidore (2002)

On September 9, a tropical wave moved off the African coast. It intensified to a hurricane late on September 19 while south of Cuba. After making landfall on Cabo Frances late on September 20, the hurricane crossed the island, then slowed as it moved westward across the Gulf of Mexico. Isidore made landfall at Telchac Puerto in Yucatán as a major hurricane on September 22. It weakened rapidly as it nearly stalled over Yucatán for 30 hours, and was only a minimal tropical storm. It then moved north and hit Grand Isle, Louisiana on September 26 as a 57-knot tropical storm, and weakened quickly into a tropical depression after moving inland. Refer to NHC's report on Isidore [21] for details.

Comparisons of modeled and measured wind speeds/wind directions at ten buoys are shown in Figure 2. The results for wind direction were quite good at all buoys. As for wind speeds, at Buoy 42001 which was very close to the hurricane track, there was an unrealistic peak around September 26, and there were some underestimations at Buoy 42007. Those discrepancies at both buoys might be caused by missing radii data. At other buoys which were relatively far from the track, such as Buoys 42036, 42039 and 42040, modeled wind speeds compared very well with the measurements. Buoys 42001, 42002, 42007, 42039 and 42040 were included in scatter plots. Although the results for wind direction were good, wind speeds scattered a lot (black and blue points), especially for observed winds larger than 20 knots (10 ms^{-1}; see blue points).

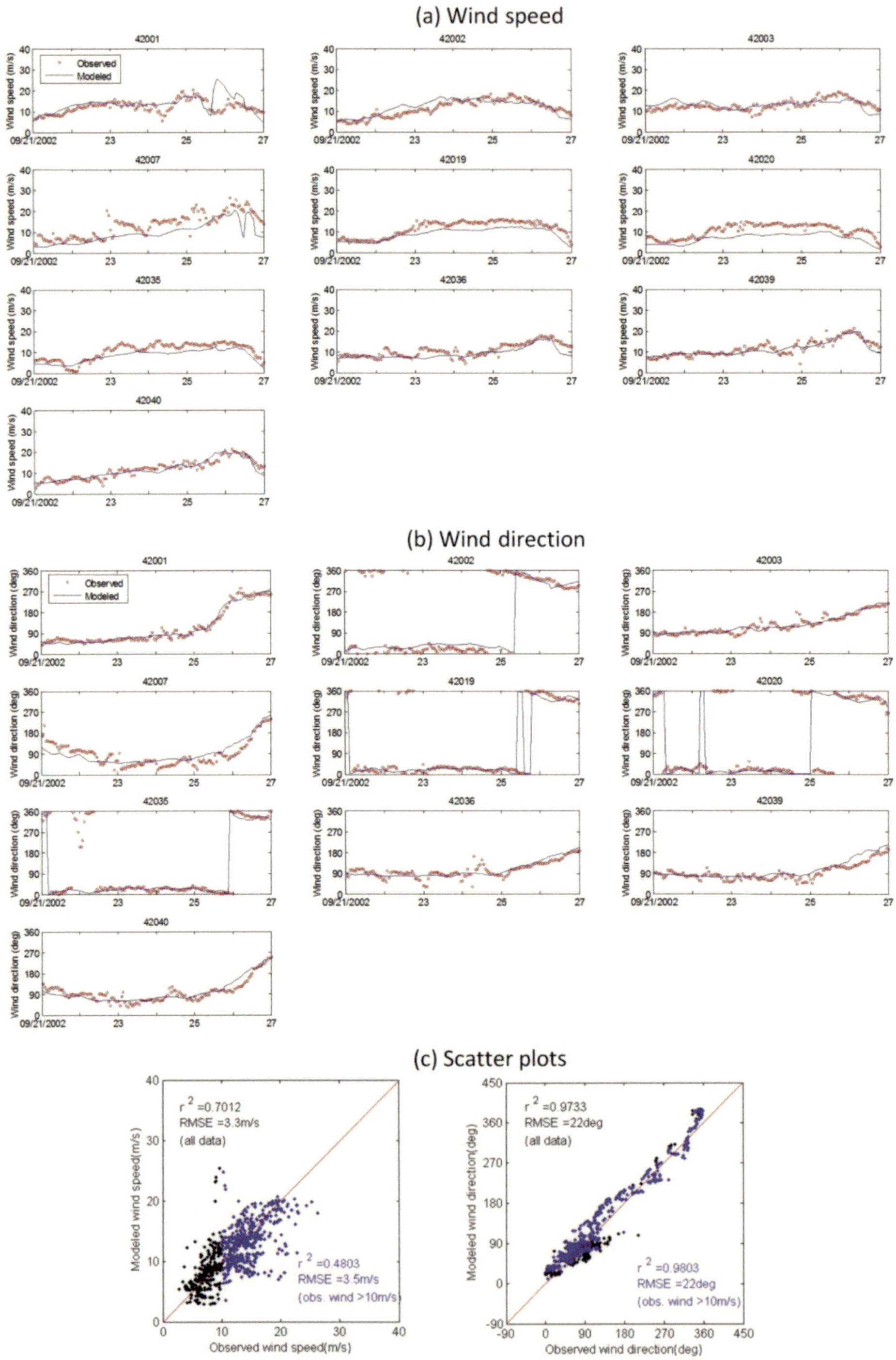

Figure 2. Wind comparisons at buoys in the Gulf of Mexico during Hurricane Isidore.

http://dx.doi.org/10.5772/51288

3.2. Hurricane Lili (2002)

Lili originated from a tropical wave that moved off the west coast of Africa on September 16. Lili became a hurricane on the 30th.

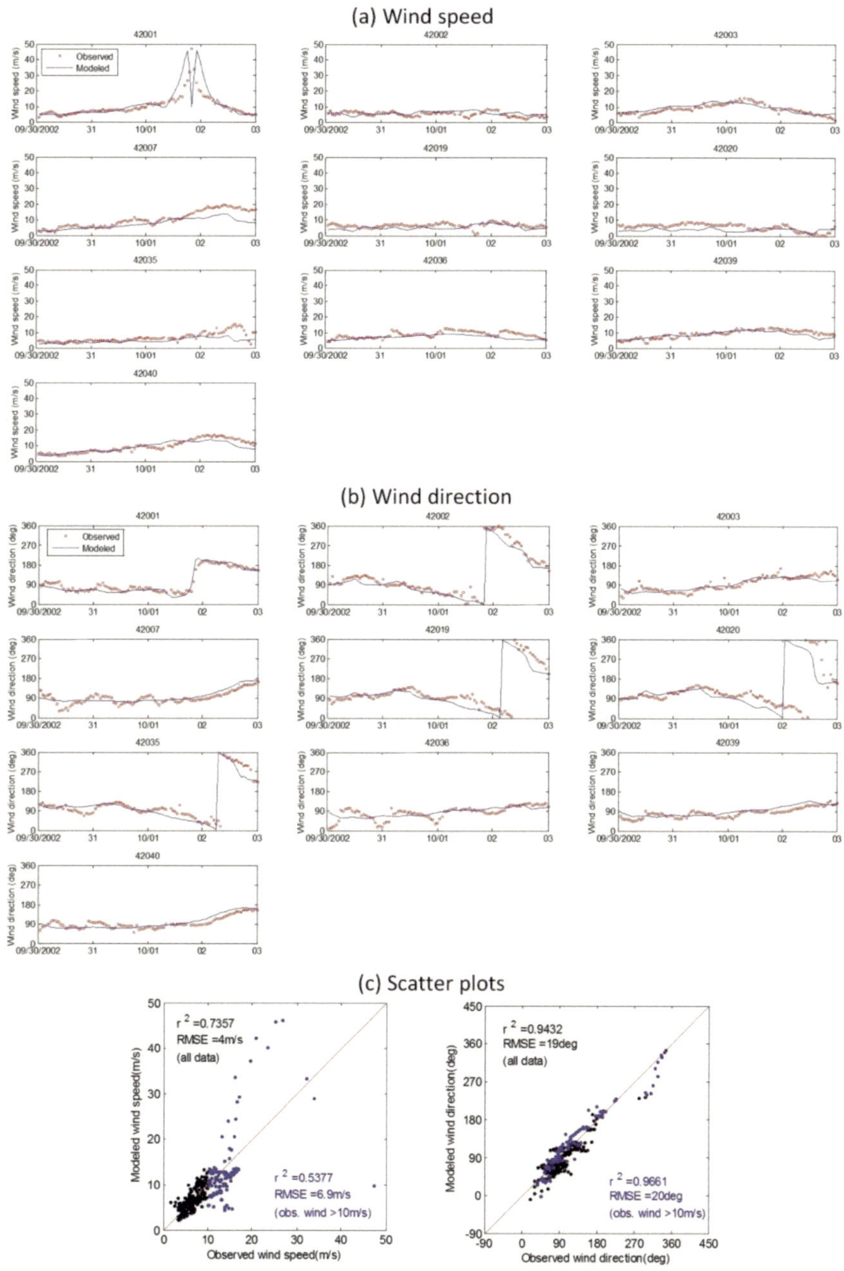

Figure 3. Wind comparisons at buoys in the Gulf of Mexico during Hurricane Lili.

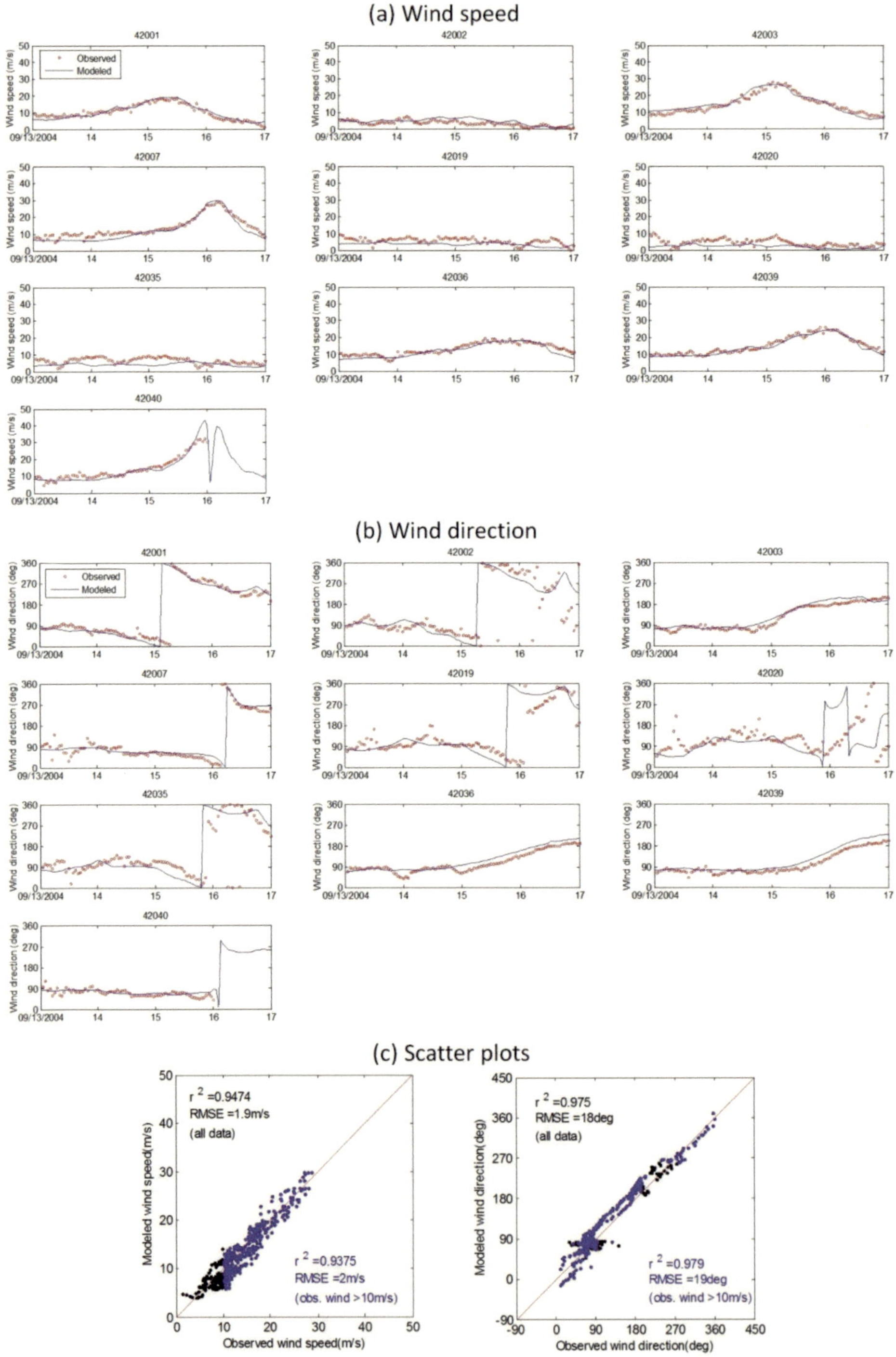

Figure 4. Wind comparisons at buoys in the Gulf of Mexico during Hurricane Ivan.

The center of the hurricane moved over the southwest tip of the Isle of Youth on the morning of October 1st. Lily turned northward and made landfall on the Louisiana coast on the 3rd, with an estimated 80-knot maximum wind speed. However, between Cuba and Louisiana, Lili intensified to 125 knots early on the 3rd over the north-central Gulf of Mexico and then rapidly weakened to 80 knots during the 13 hours until landfall. Lili was absorbed by an extratropical low on the 4th near the Tennessee/Arkansas border. Refer to NHC's report on Lili [22] for details.

Comparisons and scatter plots are shown in Figure 3. Buoys 42001, 42003, 42035 and 42040 were included in scatter plots. The results for wind direction were reasonable, while the modeled wind speeds were underestimated at Buoys 42007, 42035 and 42040. At Buoy 42001 where the hurricane almost moved through, the results showed an overestimation of wind speed. The lack of radii information and storm position error are possible explanations for the wind errors. The winds near the hurricane center change dramatically. The 6-hourly best track data cannot provide enough resolution for locations so close to the center. Scatter plots showed similar results with Isidore case.

3.3. Hurricane Ivan (2004)

Ivan was the most severe hurricane to strike the Alabama and western Florida coastlines in several decades. Ivan started as a tropical wave off the west coast of Africa on August 31, 2004. It developed into a tropical depression on September 2, 2004, in the Atlantic Ocean and strengthened into a Category 1 hurricane on the Saffir-Simpson scale 3 days later. On September 14, Ivan entered the Gulf of Mexico as a Category 5 hurricane with wind speeds of 140 knots. Ivan weakened as it moved toward the northern Gulf of Mexico and made landfall on September 16, at approximately 0650 UTC near Gulf Shores, Alabama, as a Category 3 hurricane with maximum sustained wind speeds of 113 knots. Refer to NHC's report on Ivan [23] for details.

Comparisons and scatter plots are shown in Figure 4. The results by wind model were quite good except for some overestimation of wind speed at Buoy 42040 and minor phase difference for wind direction at Buoys 42036 and 42039. Due to a malfunction at Buoy 42040 during Ivan, the recorded data for the last few hours may not be reliable. Buoys 42001, 42003, 42007, 42036 and 42039 were included in scatter plots. The RMSE values for wind speeds and wind directions are only about 2 ms^{-1} and 20 degrees.

3.4. Hurricane Dennis (2005)

Dennis was an unusually strong July major hurricane that left a trail of destruction from the Caribbean Sea to the northern coast of the Gulf of Mexico. Dennis formed from a tropical wave that moved westward from the coast of Africa on June 29. Dennis traversed a long section of western Cuba before emerging into the Gulf of Mexico on July 9. Dennis gradually intensified over the Gulf of Mexico, then rapidly intensified on July 9-10. Maximum sustained winds reached a peak of 125 knots on July 10. Thereafter, the maximum sustained winds decreased to 105 knots and the central pressure rose to 946 mb before Dennis made landfall on Santa Rosa Island, Florida, between Navarre Beach and Gulf Breeze the same day. Refer to NHC's report on Dennis [24] for details.

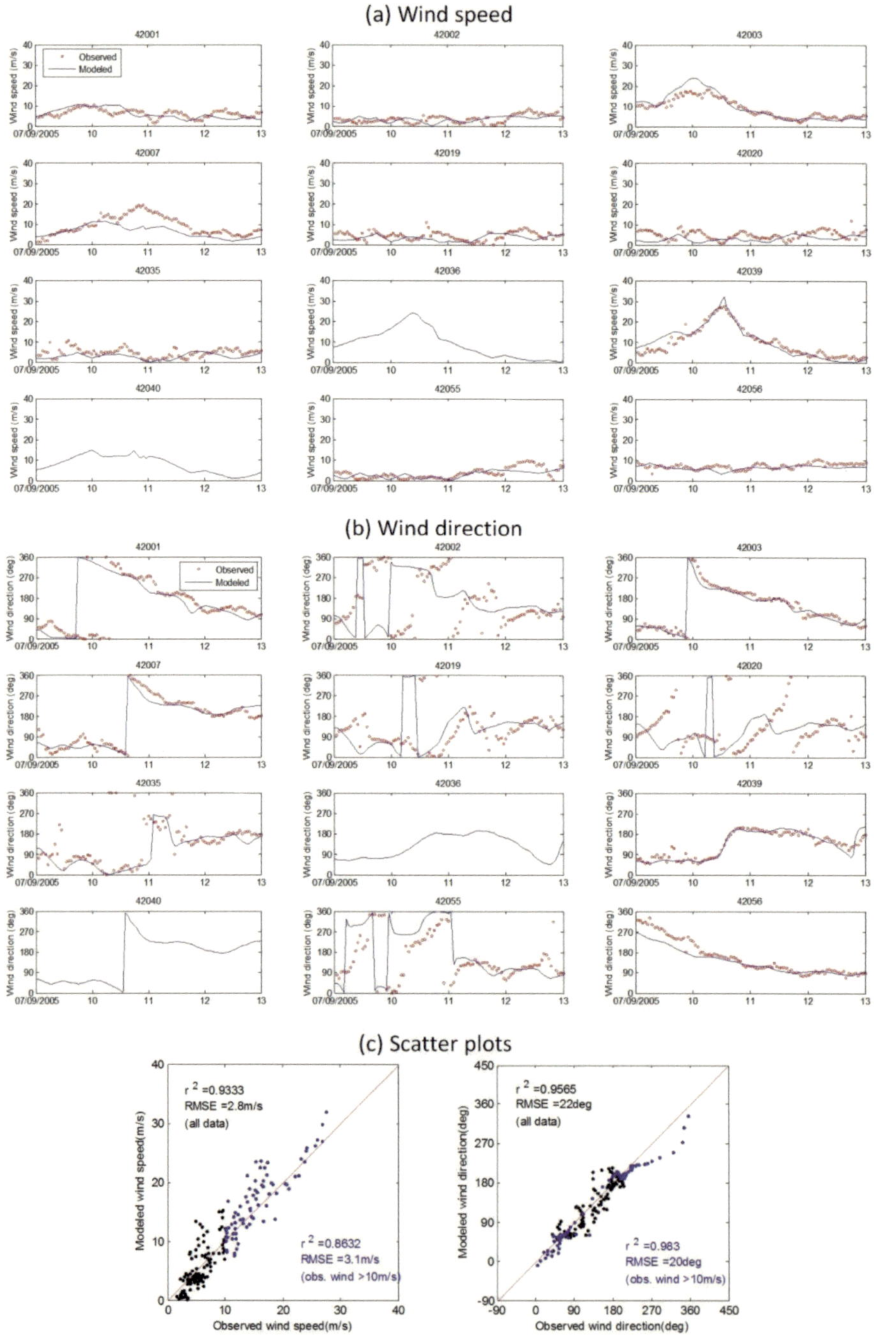

Figure 5. Wind comparisons at buoys in the Gulf of Mexico during Hurricane Dennis.

http://dx.doi.org/10.5772/51288

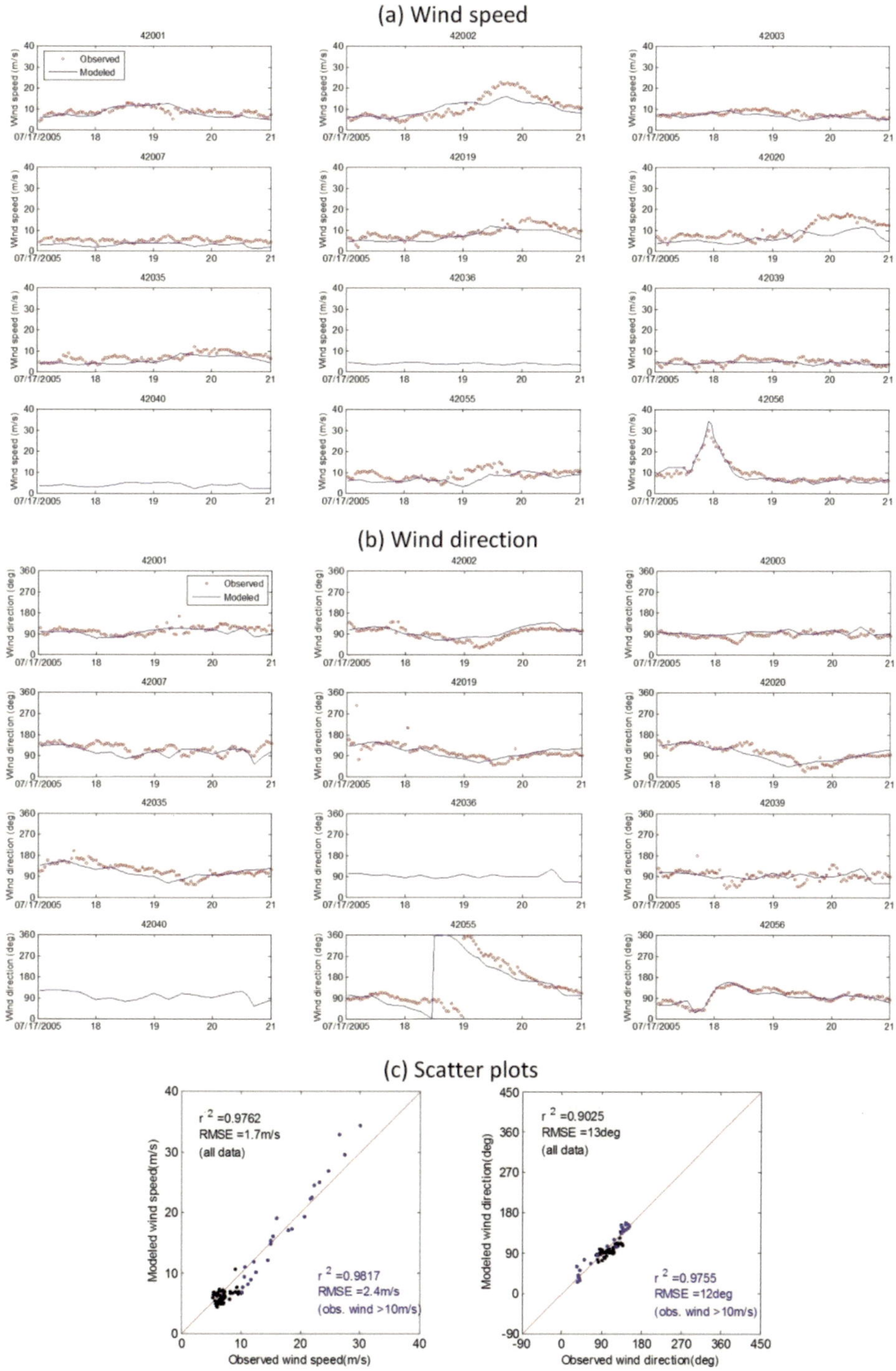

Figure 6. Wind comparisons at buoys in the Gulf of Mexico during Hurricane Emily.

Comparisons and scatter plots are shown in Figure 5. Wind speeds and wind directions at Buoy 42039 were very close to the hurricane track and our scheme validated reasonably well. There was some overestimation/underestimation of wind speed at Buoys 42003 and 42007, respectively. At other buoys where the background winds dominate, there were some errors mainly due to the operational model resolution. Buoys 42003 and 42039 were included in scatter plots. The overestimation of wind speed at Buoy 42003 increased the final RMSE to 3 ms^{-1}.

3.5. Hurricane Emily (2005)

Emily developed in the deep tropics, and by 13 July, had strengthened into a hurricane about 100 miles southeast of Grenada. Emily strengthen to a Category 5 hurricane late on 16 July at 17.1°N 79.5°W with maximum sustained surface wind speeds near 140 knots and a minimum surface pressure of 929 mb. Emily maintained hurricane intensity after crossing over into the southwestern Gulf of Mexico. Once back over open water, Emily again began intensifying, and approached extreme northeastern Mexico as a potent Category 3 hurricane with maximum sustained surface winds of 110 knots. Emily made landfall on July 20 around San Fernando, Mexico, and weakened to a tropical depression inland next day. Refer to NHC's report on Emily [25] for details.

Comparisons and scatter plots are shown in Figure 6. Similar with the situation in Dennis case, the modeled wind speeds and wind directions at Buoy 42056 which located close to the hurricane track agreed very well with the measurements. At Buoys 42002 and 42020, however, wind speeds are underestimated. Only Buoy 42056 was considered in scatter plots. The RMSE of wind directions is 12 degrees.

3.6. Hurricane Katrina (2005)

Katrina was an extraordinarily powerful and deadly hurricane that carved a wide swath of catastrophic damage and inflicted large loss of life. It was the costliest and one of the five deadliest hurricanes to ever strike the United States. Katrina started as a tropical depression on August 23, 2005, in the Atlantic Ocean. On August 28, Hurricane Katrina reached Category 5 status with wind speeds of 152 knots and a pressure of 902 mb near the center of the Gulf of Mexico. Hurricane Katrina made its second landfall between Grand Isle, Louisiana and the mouth of the Mississippi River on August 29, as a Category 3 hurricane with wind speeds of 110 knots and a low pressure of 920 mb. By 1500 UTC, Katrina made its third landfall near the Louisiana and Mississippi borderline as a Category 3 hurricane with wind speeds of nearly 105 knots. Refer to NHC's report on Katrina [26] for details.

Comparisons and scatter plots are shown in Figure 7. Buoys 42001, 42003, 42007 and 42030 were included in scatter plots. The modeled results were fairly good except for some overestimation of wind speed at Buoys 42036 and 42039. At Buoy 42040 where the winds are strongest among other buoys, the agreements of both wind speed and wind direction were excellent.

http://dx.doi.org/10.5772/51288

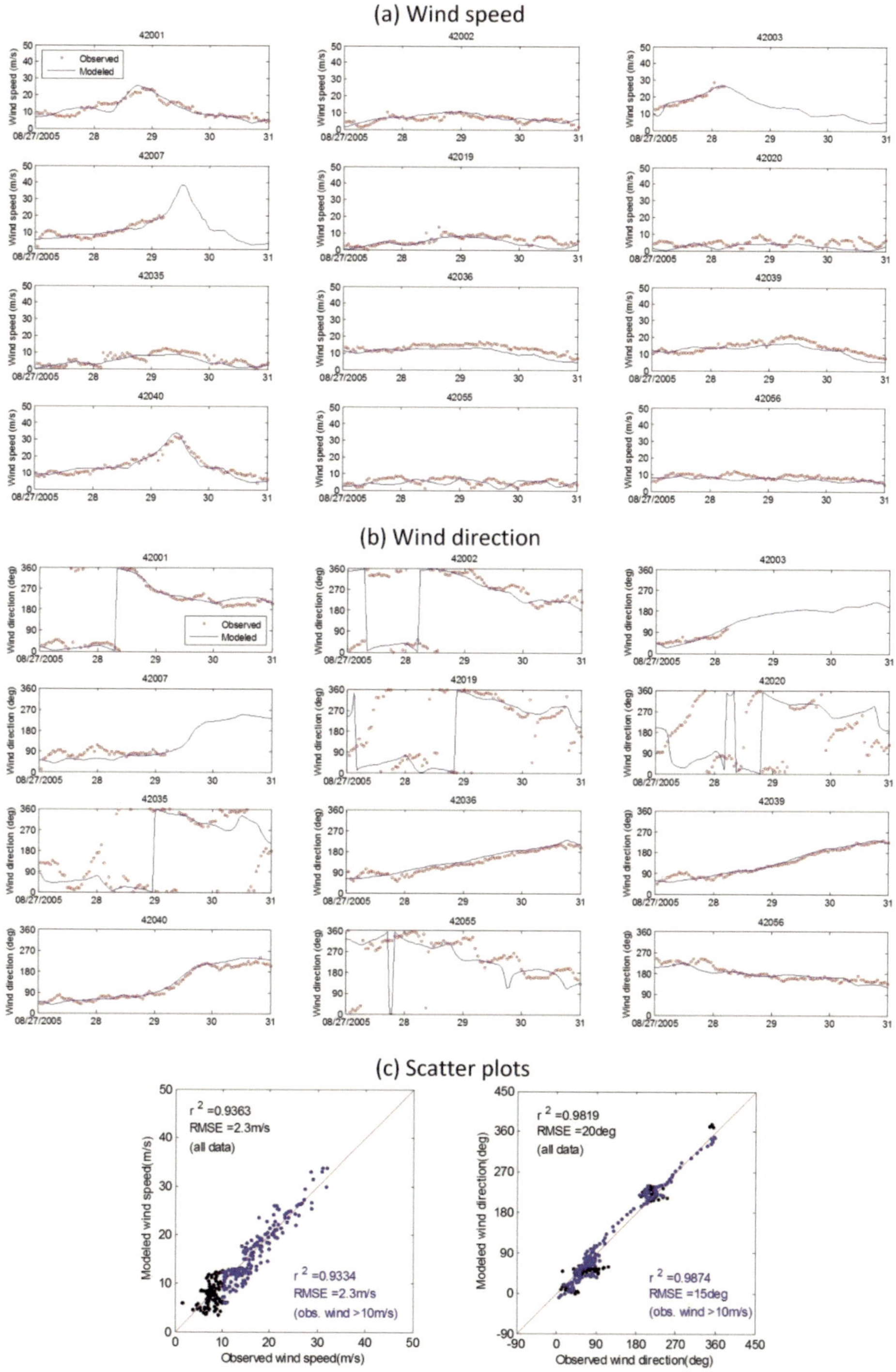

Figure 7. Wind comparisons at buoys in the Gulf of Mexico during Hurricane Katrina.

3.7. Hurricane Rita (2005)

Rita officially became a tropical storm on September 18, 2005, moved through the Florida Straits, and approached the Florida Keys on the 20th.

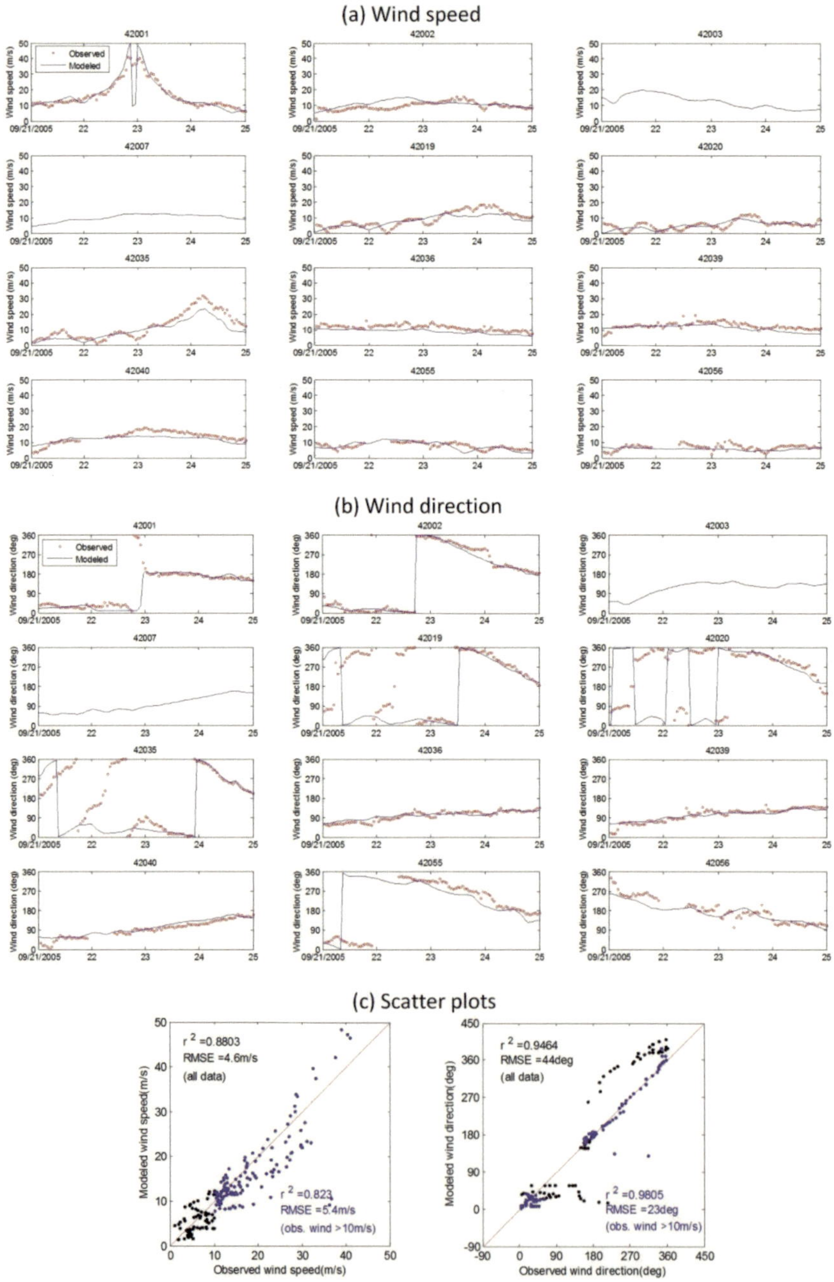

Figure 8. Wind comparisons at buoys in the Gulf of Mexico during Hurricane Rita.

http://dx.doi.org/10.5772/51288

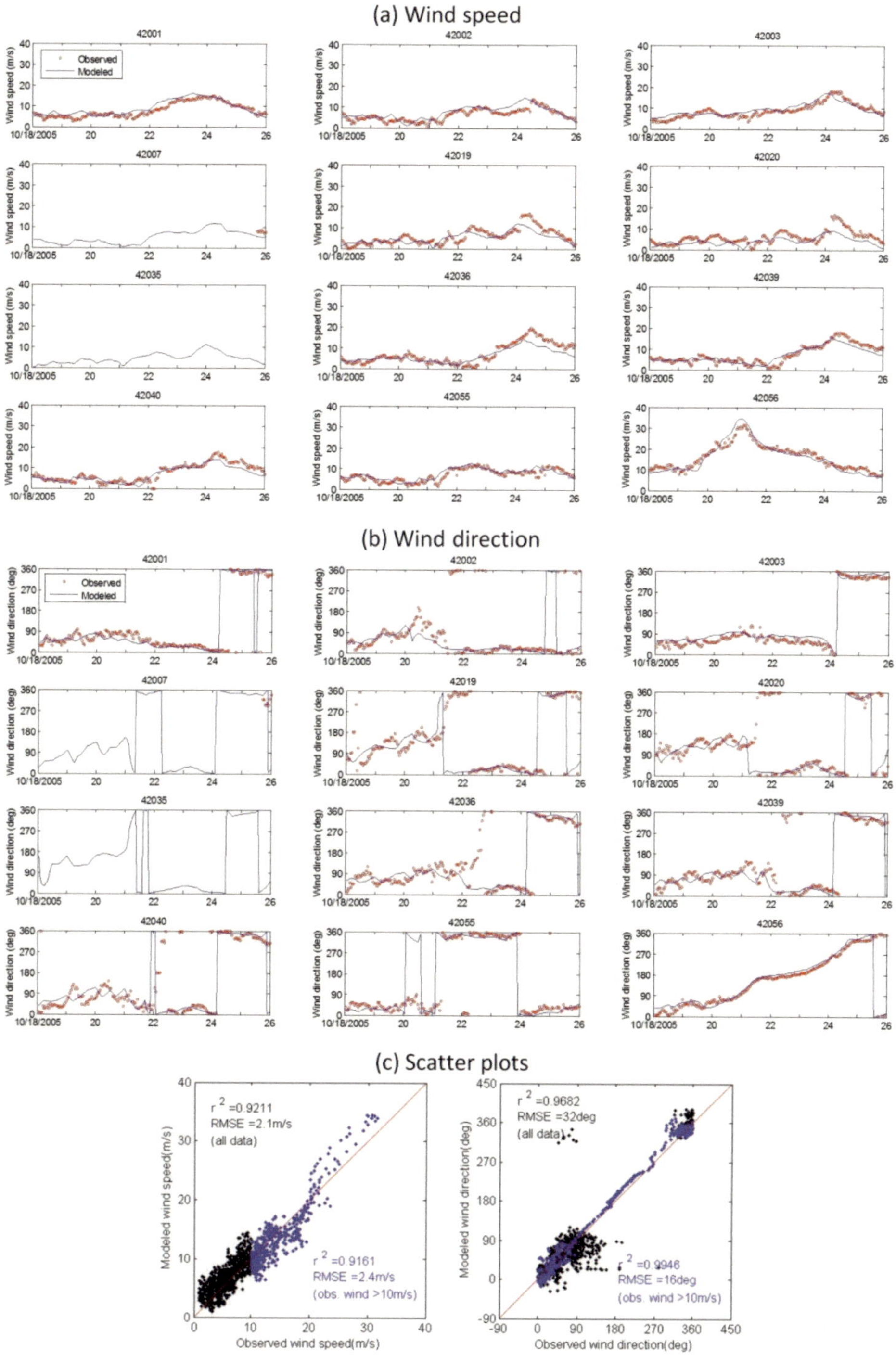

Figure 9. Wind comparisons at buoys in the Gulf of Mexico during Hurricane Wilma.

Rapidly intensifying, Rita tracked westward into the Gulf of Mexico and by the afternoon of the 21st, Rita had reached Category 5 strength, with winds of 143 knots. It peaked at 152 knots with a minimum central pressure of 897 mb. Weakening on the 22nd, Rita's intensity dipped and continued to weaken gradually over the next 36 hours prior to landfall. Rita tracked west-northwest during the 23rd and made landfall at the Texas/Louisiana border early on the 24th, at Category 3 strength with sustained winds of 104 knots. Refer to NHC's report on Rita [27] for details.

Comparisons and scatter plots are shown in Figure 8. Buoys 42001 and 42035 were included in scatter plots. Rita moved near Buoy 42001, and our scheme overestimated the peak wind speed a little bit. At Buoy 42035 before hurricane landfall, the model underestimated wind speeds, possibly due to a position error from the best track data. The scatter plots showed that the overall agreement was not very good in this case.

3.8. Hurricane Wilma (2005)

Hurricane Wilma formed in the second week of October 2005. From October 18, and through the following day, Wilma underwent explosive deepening over the open waters of the Caribbean; in a 30-hour period, the system's central atmospheric pressure dropped to the record-low value of 882 mb, while the winds increased to 161 knots. Hurricane Wilma then weakened to Category 4 status, and on October 21, it made landfall on Cozumel and on the Mexican mainland. Wilma reached the southern Gulf of Mexico before accelerating northeastward. The hurricane re-strengthened to hit Cape Romano, Florida, as a major hurricane. Wilma weakened as it quickly crossed the state, and entered the Atlantic Ocean. By October 26, it transitioned into an extratropical cyclone. Refer to NHC's report on Wilma [28] for details.

Comparisons and scatter plots are shown in Figure 9. At Buoy 52056, the hurricane winds were well reproduced. Agreements at other buoys were fairly accurate except for some underestimation of wind speed at a few buoys, e.g. 42036. Buoys 42001, 42002, 42003, 42036, 42039, 42055 and 42056 were included in scatter plots. It can be seen that the modeled wind directions agreed very well with the measurements for observed winds larger than 10 ms^{-1}.

3.9. Hurricane Gustav (2008)

Hurricane Gustav was a Category 4 hurricane and caused many deaths and considerable damage in Haiti, Cuba, and Louisiana. Gustav formed from a tropical wave that moved off the coast of Africa on August 13, 2008. Gustav strengthened into a hurricane before making landfall in Haiti. After significantly weakening over Haiti, Gustav encountered the warm waters of the northwestern Caribbean Sea, allowing for rapid intensification on August 30 before making landfall on the Isle of Youth, Cuba, and weakened to a tropical storm. Continuing into the Gulf of Mexico, Gustav regained some strength, making landfall near Cocodrie, LA as a Category 2 storm with maximum sustained winds of 90 knots. Refer to NHC's report on Gustav [29] for details.

Comparisons and scatter plots are shown in Figure 10. Buoys 42001, 42003, 42007, 42036, 42039 and 42040 were included in scatter plots. The modeled wind speeds matched the maximum wind speeds at each buoy reasonably well, although on September 2, the model over-estimated wind speeds. The agreement of wind direction was very good.

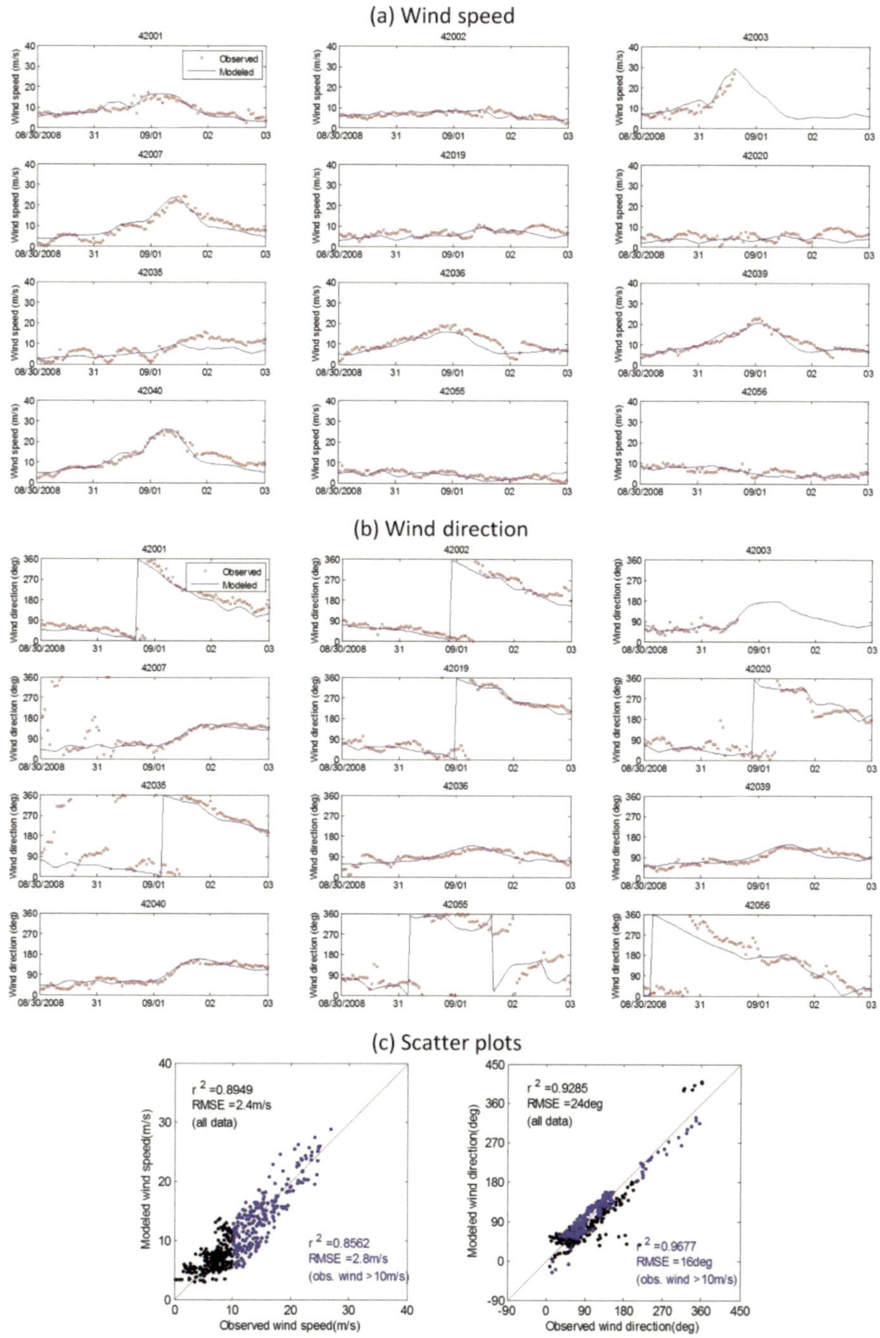

Figure 10. Wind comparisons at buoys in the Gulf of Mexico during Hurricane Gustav.

3.10. Hurricane Ike (2008)

Ike was a long-lived Cape Verde hurricane that caused extensive damage and many deaths across portions of the Caribbean and along the coasts of Texas and Louisiana.

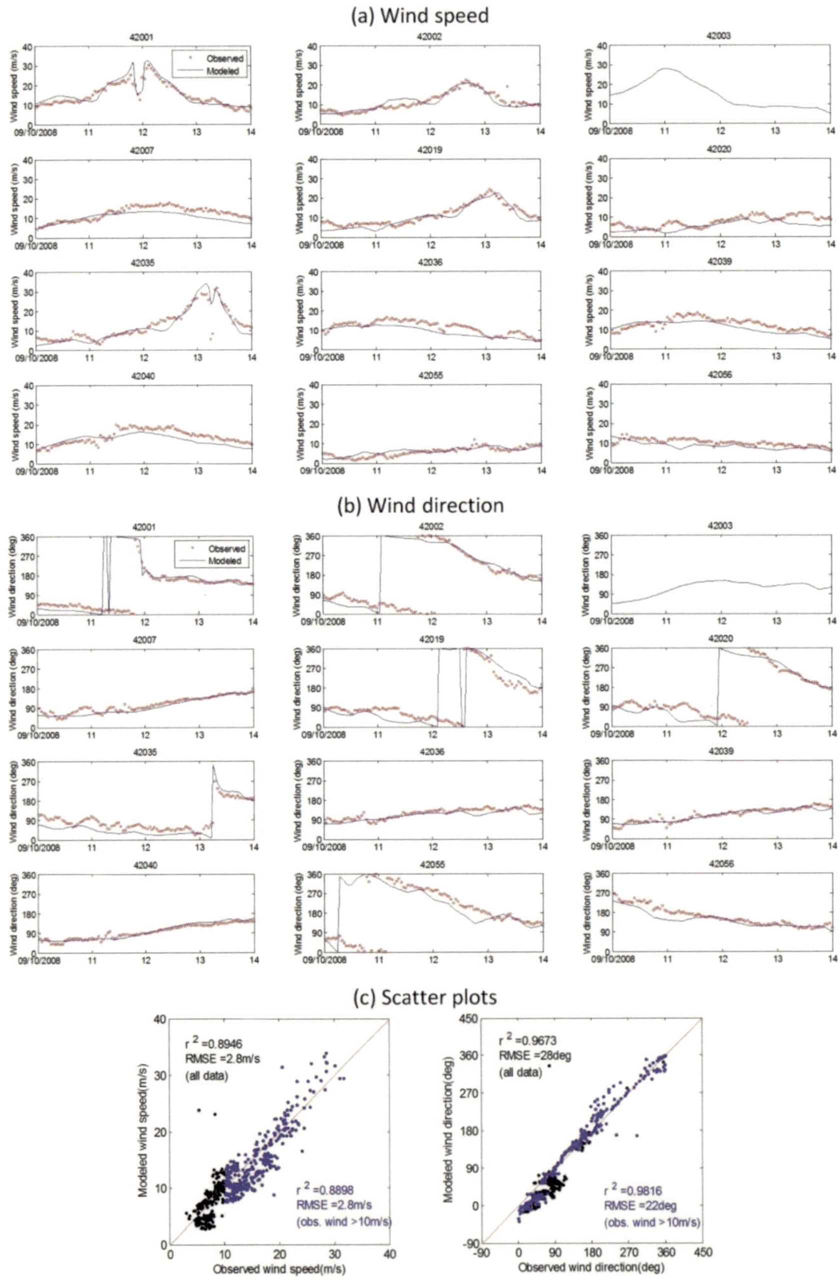

Figure 11. Wind comparisons at buoys in the Gulf of Mexico during Hurricane Ike.

Ike developed from a vigorous tropical wave that emerged off the west coast of Africa on August 29. Hurricane Ike made landfall across Great Inagua as a Category 4 hurricane on the morning of September 7, moving into the northeast Coast of Cuba as a Category 3 hurricane later that evening. Once Ike emerged into the Gulf of Mexico, the storm tracked more northwestward. Ike continued to grow on September 11. Ike continued tracking towards the upper Texas Coast, becoming better organized. Ike made landfall on Galveston Island September 13 as a strong Category 2 with sustained winds of 96 knots and a central pressure of 952 mb. Refer to NHC's report on Ike [30] for details.

Comparisons and scatter plots are shown in Figure 10. Buoys 42001, 42002, 42019, 42035 and 42040 were included in scatter plots. The wind evolution was reproduced quite well at Buoy 42001. The agreement at Buoy 42035 was also good although the calculated wind decrease was underestimated. Results at other buoys were reasonable except for some minor underestimation at Buoys 42036 and 42040 which were further away from Ike's center.

3.11. Summary

Scatter plots for eight hurricanes (except two 2002 hurricanes) at each related set of buoys are shown in Figure 12a. In addition, comparisons were made to another dataset which use the original shape parameter in Eq. (2) but restricted values between 1.0 and 2.5, used only the largest specified wind radii, and did not combine wind fields but did consider background operational model data, and is designated the "experimental wind model" (Figure 12b). The experimental model uses the highest available specified wind speed (34-, 50-, or 64-knot) and its 4-quadrant radii to generate the asymmetric hurricane wind, and the effect of the translational velocity is not excluded from those specified wind speeds. Scatter plots of observed winds versus experimental winds at same buoys for eight hurricanes are shown in Figure 12b. It can be seen that the overall RMSE of wind speeds was less than 3 ms^{-1} by using the improved parametric wind model. Without those improvements, the RMSE would increase to 3.4 ms^{-1} for observed winds larger than 10 ms^{-1}. The effect for wind directions was not as obvious as that for wind speed, but still was positive. Due to the low resolution of background winds, the RMSE of wind directions for all data was 30 degrees. When only considering larger winds (>10 ms^{-1}) which are mainly hurricane-dominated, the RMSE reduced to 19 degrees.

Among ten hurricane cases, the RMSE of modeled wind speeds for Hurricanes Isidore and Lili was relatively large, mainly due to the lack of radii information. Among the remaining eight cases, the RMSE in Rita case was larger than others. There are several factors which can cause model errors. The low resolution (uncertainty) of the location of hurricane track may induce significant changes in winds near hurricane center. Local weather processes which are not considered in the model would influence the results in some cases, especially for some nearshore buoys, e.g. 42007 and 42035. The error in best track data can ruin the results directly because the accuracy of a parametric wind model depends on the quality of the input hurricane parameters.

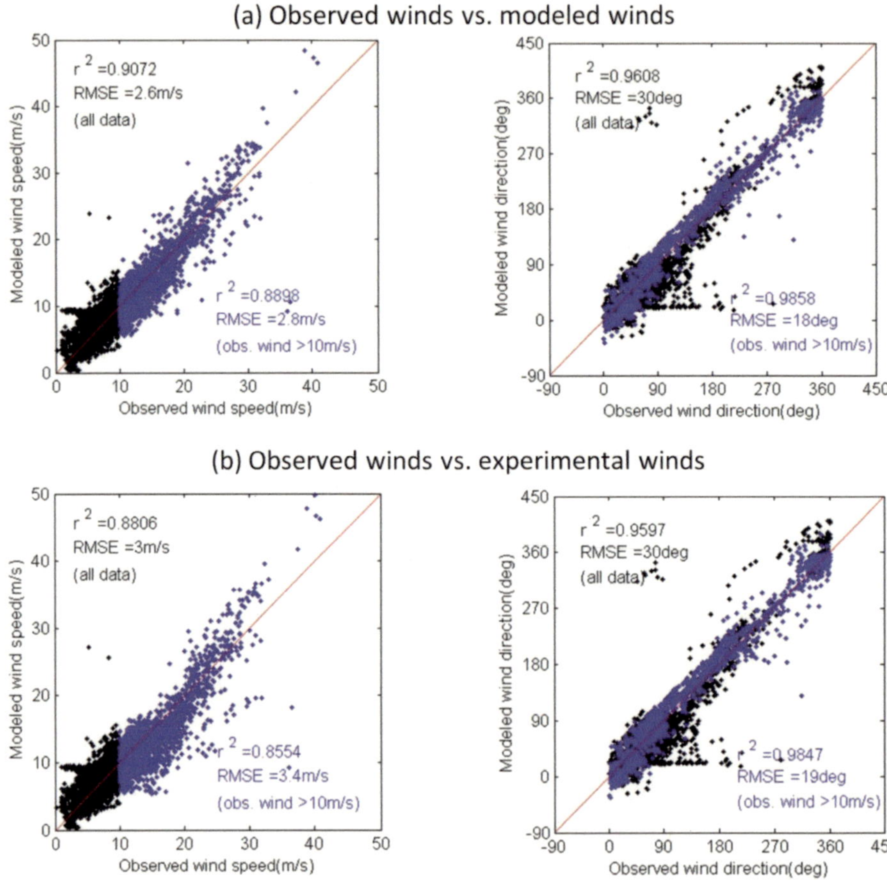

Figure 12. Scatter plots of observed winds versus (a) modeled winds and versus (b) experimental winds for eight hurricanes.

4. Discussion

4.1. Comparison with H*Wind

The H*Wind surface wind analysis [1] is considered to be one of the best hurricane wind estimates available. One question is whether this simple parametric model can reproduce the H*Wind distributions with the same data input? Two test cases for Hurricanes Katrina and Gustav were carried out. The difference from the previous cases is that input hurricane parameters were all taken from the H*Wind data except central pressure, which used best track. If applicable, the radii of all four-quadrant wind speeds (34, 50, 64 and 100 knots) were extracted from the H*Wind dataset and used in the wind model. Comparisons of wind field are shown in Figure 13. At 0000 UTC, August 29, 2005, during Katrina, the wind field produced by the wind

http://dx.doi.org/10.5772/51288

model agreed very well with the H*Wind distribution. The red circle is the extension of four points (four-quadrant radii) taken from the corresponding white circle in H*Wind. A good fitness between four red circles and four white circles in H*Wind means that the four-quadrant radii for four specified wind speeds can represent the actual wind structure in H*Wind. For Gustav at 0130 UTC, September 1, 2008, the red circles did not match the white ones well in H*Wind, which means a four-point scheme cannot capture asymmetric complexities. Otherwise, if the H*Wind asymmetry of wind structure is not very complicated, using 4-quadrant radii in this parametric wind model is capable of reproducing H*Wind distribution.

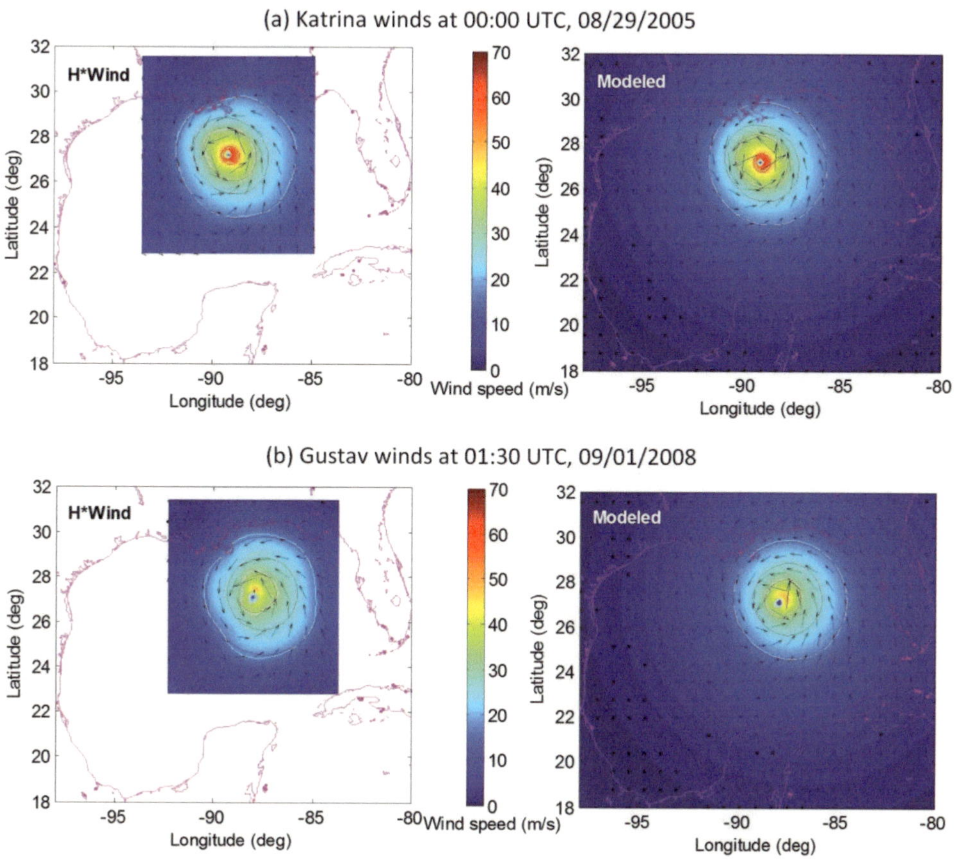

Figure 13. Comparisons of H*Wind distributions and modeled wind fields during (a) Hurricane Katrina and (b) Hurricane Gustav. Black arrows denote wind vectors. Write lines are the 34, 50, 64 and 100 knot (i.e., 17.5, 25.5, 33 and 51.5 ms^{-1}) wind speed contours. Red lines are the 34, 50, 64 and 100 knot wind speed contours fitted to the four input radii. Note that 100 knot contour is not applicable for Gustav.

Figure 14 depicts the comparison of the wind swaths generated from the H*Wind, the modeled winds and the experimental winds for Hurricanes Katrina and Gustav. In both cases, the modeled swath based on the improved parametric model agrees fairly well with the H*wind swath of maximum winds, while the swath based on the experimental winds shows some obvious discrepancies. In experimental winds, the maximum wind band on the right side of the hurricane track appears to be "broader" than that in H*Wind or modeled winds, and on the left side, the experimental result seems "narrower". This means that without those improvements in the parametric wind model, the asymmetry in wind structure would be exaggerated.

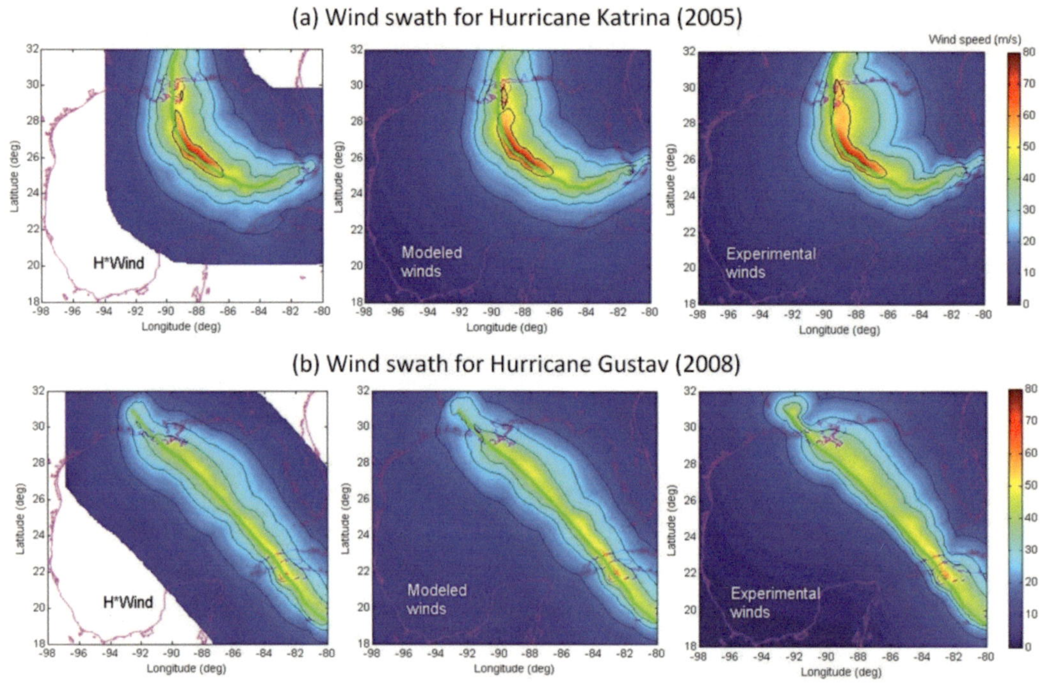

Figure 14. Wind swaths for H*Wind, modeled winds and experimental winds during (a) Hurricane Katrina and (b) Hurricane Gustav. Green line denotes hurricane track. Black lines are the 34, 50, 64 and 100 knot (i.e., 17.5, 25.5, 33 and 51.5 ms^{-1}) wind speed contours. Note that 100 knot contour is not applicable for Gustav.

4.2. Effects on surge and wave modeling

In order to further demonstrate the importance of those improvements in the parametric wind model, using Hurricane Gustav as an example, a fully coupled storm surge and wind wave modeling system [31] was employed to test the results. The ADCIRC model [32] and the SWAN model [33] are used to calculated storm surge and spectral waves, respectively. The same unstructured mesh from the Coastal Protection and Restoration Authority (CPRA) of Louisiana was adopted for both models. The CPRA mesh has about 1 million nodes and 2 million elements, covering the Gulf of Mexico and part of the Atlantic Ocean. The mesh resolution varies from 114 km in the Atlantic Ocean to about 20 m in Louisiana and Mississippi. Seven tidal constituents (M_2, S_2, N_2, K_1, O_1, K_2 and Q_1) are considered by harmonic constants at the open boundary in the Atlantic Ocean. The time steps are 1 hour and 1 second for SWAN and ADCIRC, respectively. The non-stationary mode of SWAN in spherical coordinates is used. Thirty-one exponentially spaced frequencies from 0.0314 Hz to 0.5476 Hz with 36 evenly spaced directions (10° resolution) are utilized.

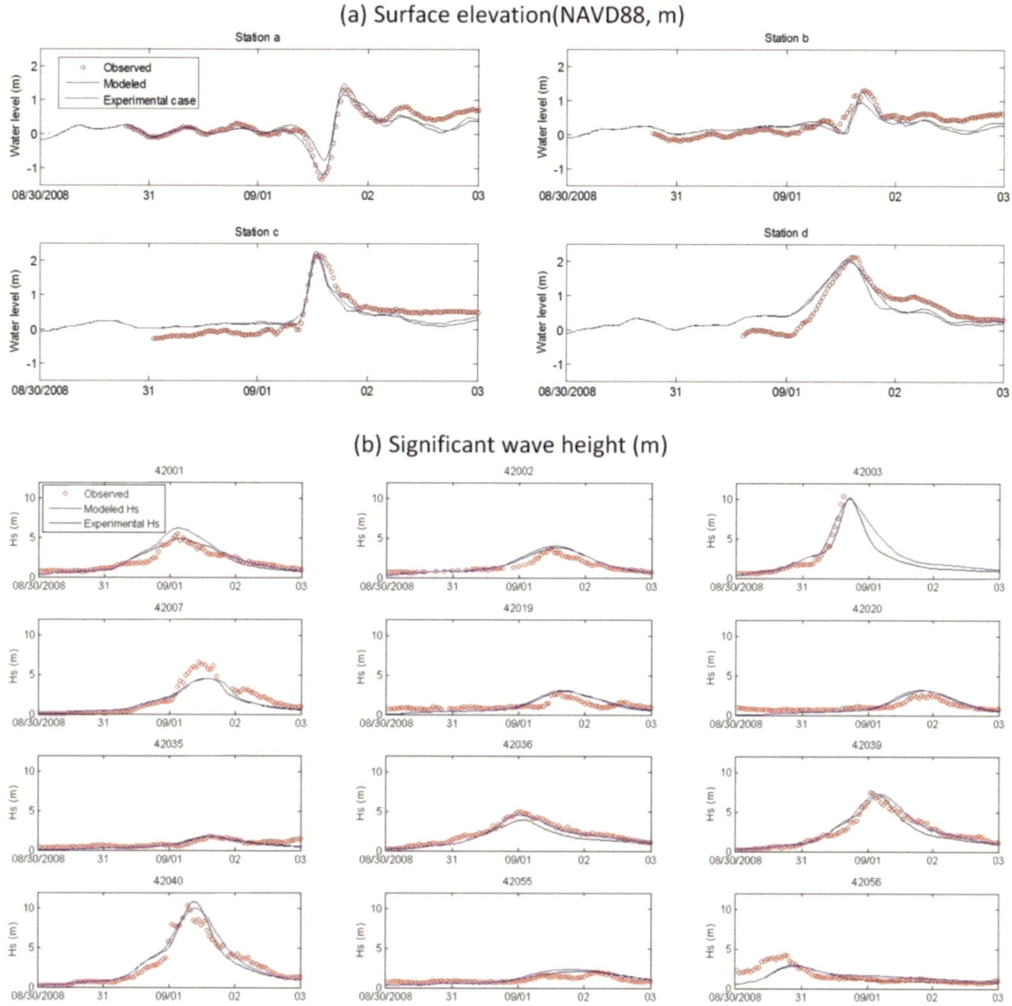

Figure 15. Comparisons of surface elevation and significant wave height using modeled winds with those using experimental winds during Hurricane Gustav.

Comparisons of modeled surge and waves with measurements are shown in Figure 15. Surge measurements were provided by some temporary gages deployed along Louisiana coast during the passage of Gustav [34]. The location of four gauges (Stations a-d) used in this paper are shown in Figure 1. Figure 15 shows an improvement of modeled results with measurements than the results driven by the experimental winds. At Station a, the modeled results reproduce the process of surge setup and set down as Gustav passed to the right of the station. At Stations a and b, located at the left side of the track, the modeled maximum surge was larger than that by the experimental winds, while at Stations c and d, located on the right side of the track, the experimental case shows a surge higher than observed. Wave height comparisons showed that the modeled wave height agreed very well with the measurements, e.g. at Buoy 42036, except for some overestimation at Buoy 42001 and some underestimation at Buoy 42007. The experimental winds result in higher wave heights at Buoy 42040 located to the right side of the track, and lower wave heights at Buoy 42001 located to the left side. This is consistent with the swath results, narrower at the left side and broader at the right side.

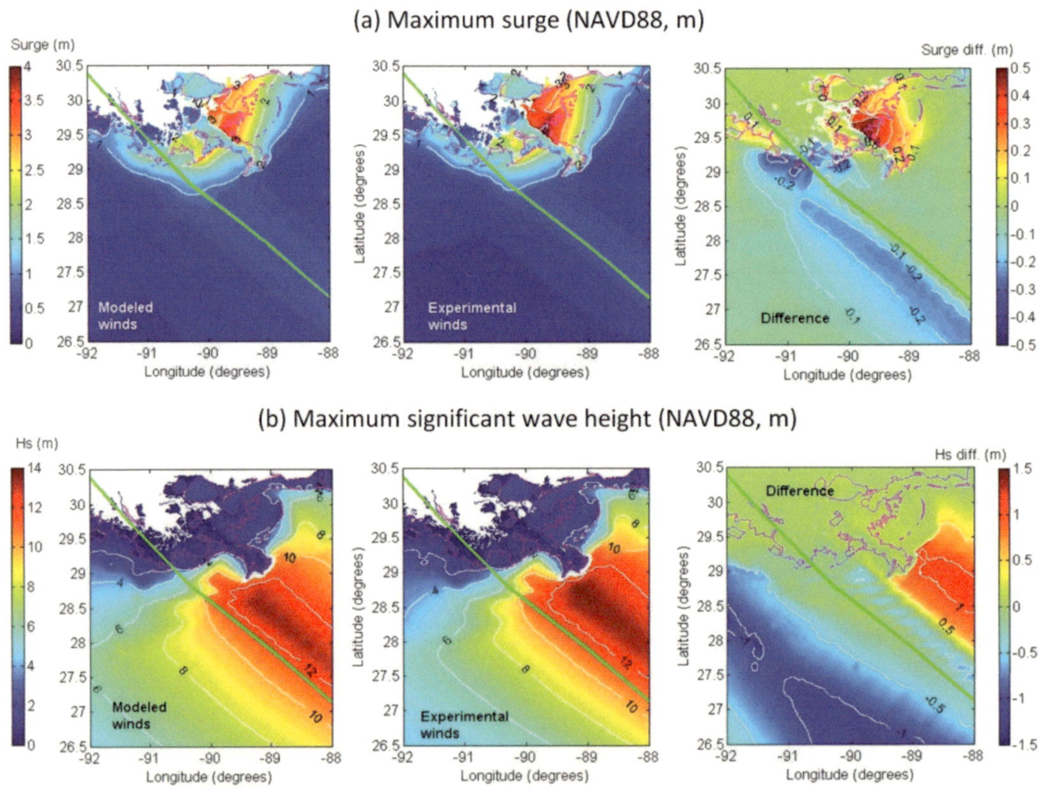

Figure 16. Comparisons of maximum surge and maximum significant wave height using modeled winds with those using experimental winds during Hurricane Gustav. Green line denotes hurricane track.

Distributions of maximum surge and maximum significant wave height driven by the modeled winds and by the experimental winds are compared in Figure 16. Difference distributions are provided as well to quantify the influence. During Gustav, counterclockwise winds pushed the water into the Chandeleur Sound and blocked by the east bank of the Mississippi River, which caused more than 3 m of storm surge. The maximum wave height was about 14 m off shore at the right side of the hurricane track. Due to the exaggeration of wind asymmetry in the experimental winds, compared to the modelled wind results, the maximum surge at the right bank of the Mississippi River increased about 0.5 m, and the surge along the left side of the track decreased about 0.2 m. Simultaneously, the maximum wave height at the right side of the track increased up to 1.2 m, while at the left side decreased by more than 1 m. It is shown that the accuracy of hurricane winds is of great importance to the results of storm surge and waves. Without the improvements in the parametric wind model, the difference would be about 10% on both sides of the hurricane track.

5. Conclusions

Ten historical hurricanes were selected to test the performance of a revised parametric wind model. The revisions include: retention of the Coriolis effect in the shape parameter B; translational velocity excluded before applying the Holland-type vortex to avoid exaggeration of the wind asymmetry but added back in at the end of the procedure; and a weighted composite wind field that makes full use of all wind parameters, not just the largest available specified wind speed and its 4-quadrant radii. This scheme was validated against buoy and tide gauge observations. The overall RMSE for wind speed and wind direction is 2.8 ms^{-1} and 18 degrees for observed winds larger than 10 ms^{-1}. Without those improvements, the two values of RMSE increase to 3.4 ms^{-1} and 19 degrees for another dataset (designated the "experimental" dataset) which used the original shape parameter in Eq. (2) but restricted values between 1.0 and 2.5, used only the largest specified wind radii, and did not combine wind fields but did consider background operational model data.

The low resolution of hurricane locations, local weather processes, and error in best track data may affect the accuracy of the wind model. By comparing with H*Wind, it is found that the wind model has the capability to reproduce the H*Wind distribution if the asymmetry of wind structure is not too complicated. The wind swath results produced by the experimental winds showed that the asymmetry in wind structure was exaggerated. The importance of those improvements in the parametric wind model were further demonstrated through the effects on storm surge and hurricane wave modeling by using a fully coupled surge-wave model during Hurricane Gustav. The modeled results gave better agreements with observed surges and waves than the experimental results. Due to the exaggeration of wind asymmetry in experimental winds, the difference in storm surge and hurricane waves was about 10% on both sides of the hurricane track.

It should be noted that the number of parameters in the hurricane forecasts affects the accuracy of the parametric wind model. If more information — either the 4-quadrant radii of additional specified wind speeds (e.g., 100 knots) or more radii (e.g. at eight directions instead of only at 4 quadrants) for each specified wind speed — was available, the improved parametric model can deal with more complicated asymmetry in wind structure and produce more accurate wind fields for major hurricanes.

Acknowledgements

The study has been supported in part by the U.S. National Science Foundation (NSF) (Grant No. 0652859), the NSF Northern Gulf Coastal Hazards Collaboratory (Grant No. 1010640) and the U.S. National Oceanic and Atmospheric Administration (NOAA) through the Northern Gulf Institute (Grant No. 09-NGI-08). Computational resources were provided by the Louisiana Optical Network Initiative (LONI) and Louisiana State University.

Author details

Kelin Hu[1], Qin Chen[2*] and Patrick Fitzpatrick[3]

*Address all correspondence to: qchen@lsu.edu

1 Department of Civil and Environmental Engineering, Louisiana State University, Baton Rouge, USA

2 Department of Civil and Environmental Engineering, and Center for Computation and Technology, Louisiana State University, Baton Rouge, USA

3 Geosystems Research Institute, Mississippi State University, Stennis Space Center, USA

References

[1] Powell, M. D., Houston, S. H., Amat, L. R., & Morisseau-Leroy, N. (1998). The HRD real-time hurricane wind analysis system. *Journal of Wind Engineering and Industrial Aerodynamics*, 77 & 78, 53-64.

[2] Thompson, E. F., & Cardone, V. J. (1996). Practical modeling of hurricane surface wind fields. *Journal of Waterway, Port, Coastal, and Ocean Engineering*, 122(4), 195-205.

[3] Vickery, P. J., Skerlj, P. F., Steckley, A. C., & Twisdale, L. A. (2000). Hurricane wind field model for use in hurricane simulations. *Journal of Structural Engineering*, 126, 1203-1221.

[4] Cox, A. T., Greenwood, J. A., Cardone, V. J., & Swail, V. R. (1995). An interactive objective kinematic analysis system. Banff, Alberta, Canada. *In: Proc Fourth Int Workshop on Wave Hindcasting and Forecasting*, Atmospheric Environment Service, 109-118.

[5] Vickery, P. J., & Skerlj, P. F. (2005). Hurricane gust factors revisited. *Journal of Structural Engineering*, 131, 825-832.

[6] Holland, G. J., Belanger, J. I., & Fritz, A. (2010). A revised model for radial profiles of hurricane winds. *Monthly Weather Review*, 138, 4393-4401.

[7] Vickery, P. J., & Twisdale, L. A. (1995). Wind-field and filling models for hurricane wind-speed predictions. *Journal of Structural Engineering*, 121, 1700-1709.

[8] Schloemer, R. W. (1954). Analysis and synthesis of hurricane wind patterns over Lake Okeechobee, *NOAA Hydrometeorology Report 31*, Department of Commerce and U.S. Army Corps of Engineers, U.S. Weather Bureau, Washington, D.C., 49.

[9] Leslie, L. M., & Holland, G. J. (1995). On the bogussing of tropical cyclones in numerical models: A comparison of vortex profiles. *Meteorology and Atmospheric Physics*, 56, 101-110.

[10] Jelesnianski, C. P. (1966). Numerical computations of storm surges without bottom stress. *Monthly Weather Review*, 94, 379-394.

[11] Holland, G. J. (1980). An analytic model of the wind and pressure profiles in hurricanes. *Monthly Weather Review*, 108, 1212-1218.

[12] Hu, K., Chen, Q., & Kimball, S. K. (2012). Consistency in hurricane surface wind forecasting: an improved parametric model. *Nature Hazards*, 61, 1029-1050.

[13] Willoughby, H. E., & Rahn, M. E. (2004). Parametric representation of the primary hurricane vortex. Part I: Observations and evaluation of the Holland (1980) model. *Monthly Weather Review*, 132, 3033-3048.

[14] Willoughby, H. E., Darling, R. W. R., & Rahn, M. E. (2006). Parametric representation of the primary Hurricane Vortex. Part II: A new family of sectionally continuous profiles. *Monthly Weather Review*, 134, 1112-1120.

[15] Holland, G. J. (2008). A revised hurricane pressure-wind model. *Monthly Weather Review*, 136, 3432-3445.

[16] Levinson, D. H., Vickery, P. J., & Resio, D. T. (2010). A review of the climatological characteristics of landfalling Gulf hurricanes for wind, wave, and surge hazard estimation. *Ocean Engineering*, 37, 13-25.

[17] Vickery, P. J., Masters, F. J., Powell, M. D., & Wadhera, D. (2009). Hurricane hazard modeling: The past, present, and future. *Journal of Wind Engineering and Industrial Aerodynamics*, 97, 392-405.

[18] Xie, L., Bao, S., Pietrafesa, L. J., Foley, K., & Fuentes, M. (2006). A real-time hurricane surface wind forecasting model: Formulation and verification. *Monthly Weather Review*, 134, 1355-1370.

[19] Mattocks, C., & Forbes, C. (2008). A real-time, event-triggered storm surge forecasting system for the state of North Carolina. *Ocean Modelling*, 25, 95-119.

[20] National Oceanic and Atmospheric Administration, National Weather Service, National Centers for Environmental Prediction, National Hurricane Center. *Automated Tropical Cyclone Forecast (ATCF) best track data*, ftp://ftp.tpc.ncep.noaa.gov/atcf/archive/.

[21] Avila, L. A. (2002). *Hurricane Isidore: 14-27 September 2002. Tropical Cyclone Report*, National Oceanic and Atmospheric Administration, National Weather Service, Tropical Prediction Center, Miami, Florida.

[22] Lawrence, M. B. (2002). *Hurricane Lili: 21 September-04 October 2002. Tropical Cyclone Report*, National Oceanic and Atmospheric Administration, National Weather Service, Tropical Prediction Center, Miami, Florida.

[23] Stewart, S. R. (2004). *Hurricane Ivan: 2-24 September 2004. Tropical Cyclone Report*, National Oceanic and Atmospheric Administration, National Weather Service, Tropical Prediction Center, Miami, Florida.

[24] Beven, J. (2005). *Hurricane Dennis: 4-13 July 2005. Tropical Cyclone Report*, National Oceanic and Atmospheric Administration, National Weather Service, Tropical Prediction Center, Miami, Florida.

[25] Franklin, J. L., & Brown, D. P. (2005). *Hurricane Emily: 11-21 July 2005. Tropical Cyclone Report*, National Oceanic and Atmospheric Administration, National Weather Service, Tropical Prediction Center, Miami, Florida.

[26] Knabb, R. D., Rhome, J. R., & Brown, D. P. (2005). *Hurricane Katrina: 23-30 August 2005. Tropical Cyclone Report*, National Oceanic and Atmospheric Administration, National Weather Service, Tropical Prediction Center, Miami, Florida.

[27] Knabb, R. D., Brown, D. P., & Rhome, J. R. (2006). *Hurricane Rita: 18-26 September 2005. Tropical Cyclone Report*, National Oceanic and Atmospheric Administration, National Weather Service, Tropical Prediction Center, Miami, Florida.

[28] Pasch, R. J., Blake, E. S., Cobb, H. D., III, & Roberts, D. P. (2006). *Hurricane Wilma: 15-25 October 2005. Tropical Cyclone Report*, National Oceanic and Atmospheric Administration, National Weather Service, Tropical Prediction Center, Miami, Florida.

[29] Beven, J. L., & Kimberlain, T. B. (2009). *Hurricane Gustav: 25 August-4 September 2008. Tropical Cyclone Report*, National Oceanic and Atmospheric Administration, National Weather Service, Tropical Prediction Center, Miami, Florida.

[30] Berg, R. (2009). *Hurricane Ike: 1-14 September 2008. Tropical Cyclone Report*, National Oceanic and Atmospheric Administration, National Weather Service, Tropical Prediction Center, Miami, Florida.

[31] Westerink, J. J., Luettich, R. A., Feyen, J. C., Atikinson, J. H., Dawson, C., Roberts, H. J., Powell, M. D., Dunion, J. P., Kubatko, E. J., & Pourtaheri, H. (2008). A basin- to channel- scale unstructured grid hurricane storm surge model applied to Southern Louisiana. *Monthly Weather Review*, 136, 833-864.

[32] Luettich, R. A., Westerink, J. J., & Scheffner, N. W. (1992). ADCIRC: An advanced three-dimensional circulation model for shelves, coasts and estuaries. *Report 1: Theory and Methodology of ADCIRC-2DDI & ADCIRC-3DL. Technical Report, DRP-92-6*, U.S. Army Corps of Engineers.

[33] Booij, N., Ris, R. C., & Holthuijsen, L. H. (1999). A third-generation wave model for coastal regions:1. Model description and validation. *Journal of Geophysical Research*, 104(C4), 7649-7666.

[34] Kennedy, A. B., Gravois, U., Zachry, B., Luettich, R., Whipple, T., Weaver, R., Reynolds-Fleming, J., Chen, Q. J., & Avissar, R. (2010). Rapidly installed temporary gauging for hurricane waves and surge, and application to Hurricane Gustav. *Continental Shelf Research*, 30(16), 1743-1752.

Section 2

Meteorology

Chapter 4

The Variations of Atmospheric Variables Recorded at Xisha Station in the South China Sea During Tropical Cyclone Passages

Dongxiao Wang, Jian Li, Lei Yang and Yunkai He

Additional information is available at the end of the chapter

http://dx.doi.org/10.5772/50897

1. Introduction

The South China Sea (SCS), a large semi-enclosed marginal sea in the Western Pacific Ocean, is a region where tropical cyclones (TCs) frequently occur. As a large population lives along the coastal area, it is of great interest to understand TC behavior in the SCS.

TC, which is a tropical system, strengthens when water evaporated from the ocean is released as the saturated air rises, resulting in condensation of water vapor contained in the moist air. The movement of a TC is mainly steered by the surrounding environmental flow in the troposphere and modified by the beta-effect. TC could intensity during the movement, if the conditions (both atmospheric and oceanic) along the TC track remain favorable. TC movements in the Western North Pacific (WNP) are likely caused to a large extent by changes in the planetary-scale atmospheric circulation and thermodynamic structure associated with the El Niño phenomenon [1]. Recent studies indicate that the total number of TCs and number of TCs entering the SCS from the WNP are below normal in El Niño events but above normal during La Niña events [2]. However, for TCs formed inside the SCS, the difference in numbers between the two phases of the El Niño-Southern Oscillation (ENSO) is not as obvious. Other studies relate TC genesis to the increase in accumulated cyclone energy or potential intensity [3-4]. The monsoonal flow often exhibits a life cycle of several weeks [5], connected to the Madden–Julian oscillation (MJO), which may cause TC tracks to vary intraseasonally as well.

During TC passages, atmospheric variables, as well as air-sea condition experience significant changes. The impact of a TC on a local area can be assessed through several ways, such as satellite observation, model simulations and records from weather station. Reference [6] first used imagery from meteorological satellites to estimate TC intensities. Using remote sensing data, reference [7] suggested that the largest increase of the number and proportion of hurricanes occurred in the North Pacific, Indian, and Southwest Pacific Oceans in an environment of increasing sea surface temperature. Based on multiple satellite data and satellite-based wind retrieval techniques, reference [8] objectively estimate the surface wind fields associated with TCs and construct the Multiplatform Tropical Cyclone Surface Wind Analysis (MTCSWA) product in preparing their forecasts and advisories. Model simulation can be useful in estimating regional (short/long term) climate change. Reference [9] simulated Typhoon Leo (1999) using two nested domains in relatively coarse resolution (54 km and 18 km). MM5 was applied to simulate several characteristics of Typhoon Fitow (2001), including land falling, center position, and precipitation [10]. MM5 incorporating 4D variational data assimilation system with a full-physics adjoint model was found to greatly improve typhoon forecast in track, intensity, and landfall position [11]. A successful simulation of super typhoon Tip (1979) over the northwest Pacific was simulated by using a Limited Area Model with a horizontal resolution of 0.46875°in latitude and longitude [12]. Model parameters were investigated in simulating Typhoon Chanchu (2006) by comparing with satellite measurements [13].

Compared with satellite and model simulation, it is more difficult to realistically assess the impact of a TC on a local spot (in the ocean) from weather stations due to the scarcity of station in the ocean. However, station records, though only few, are very useful in validating model simulation and satellite measurement. Using many weather stations in the Philippines for the period of 1902 to 2005, reference [14] captured the behaviors of typhoons in its vicinity. Because the environment in mountain is more sensitive than in urban, reference [15] used the data from weather station at mountain Ali to find the trend of typhoons from short to long period. They also found that air temperature and rainfall have their own trend.

This study makes full use of the precious data recorded at Xisha station to investigate the impact of TC while passing around the station. Data and introduced in section 2. The comparisons between wind and SST observations at Xisha with remote sensing measurements are given in section 3. The main analysis of the atmospheric variables at Xisha during TC passages are shown in section 4. Conclusions and discussions are then given in section 5.

2. Data and methods

Xisha station is located at the Northern part of Xisha Islands in the SCS, on which an automatic weather station (AWS) was installed and began to operate on April 6, 2008. It collects several meteorological parameters, including wind speed, wind direction, humidity, barometric pressure, and air temperature at 10 meters above sea surface every two minutes. From April 2008 to June 2009 (Stage 1), the wind direction data were classified as 22.5° sectors (azimuths format). Since July 2009 (Stage 2), the wind direction data format changed to an accuracy of one degree, and rain sensor was mounted on the station.

Wind data at Xisha station during the period from April 2008 to November 2009 is used to validate QuikSCAT and ASCAT measurements [16]. During the period, the crossing paths

http://dx.doi.org/10.5772/50897

of both QuikSCAT and ASCAT are 5-6 km away from AWS Station, which is at 16º50′N, 112º20′E. Temperature at 2 m below sea surface was collected every 10 minutes from March 13, 2010 to August 31, 2010 by a thermometer mounted on a moored surface buoy platform, which is located north of the AWS at 16º 51′N, 112º 19′E.

QuikSCAT derived surface wind vectors are acquired from the Asia-Pacific Data-Research Center (APDRC) in Near Real Time format (NRT) [17]. The data set consists of globally gridded values (0.25°×0.25°) of zonal and meridional wind velocity components at 10 m over the sea surface measured twice daily from April 6, 2008 to November 19, 2009. The wind measurement has the accuracy of 2 m/s for speed and 20° for direction for both ascending and descending passes. Since there is no QuikSCAT data after October 2009, ASCAT Ocean Surface Wind Vectors data of 25-km resolution production has also been chosen in its NRT format from March 13, 2010 to August 31, 2010. The ASCAT wind data at 10 m above the sea surface was processed by the National Oceanic and Atmospheric Administration (NOAA) utilizing measurements from ASCAT aboard the EUMETSAT METOP satellite.

Before and after the onset of the SCS summer monsoon, the SCS suffers from high fraction cloud coverage, and the Advanced Microwave Scanning Radiometer for the Earth Observing System (AMSR-E) has capability to see through clouds. Therefore, sea surface temperature data derived from AMSR-E is applied here.

Best-track dataset including TC location and intensity at six-hourly intervals produced by the Joint Typhoon Warning Center (JTWC) is used in the study. The best-track archives contain six-hourly TC positions and Maximum sustained wind speed in knots as intensity estimates. Environmental variables are derived from the U.S.

National Centers for Environmental Predication (NCEP) reanalysis [18] and 850,700,600,500,400,300,250,200 (mb) level wind data were combined as the steering wind to analysis TC environmental field.

In order to show atmospheric environmental variations with high common power and the phase relationship [19], Morlet wavelet transform, that is a continuous function, is used to evaluate the cross wavelet of two time-series x_n and y_n. The cross wavelet transform (XWT) is defined as following: $|W_{xy}| = abs(\frac{|W_x W_y^*|}{\sigma_x \sigma_y})$, Where * denotes complex conjugation, and σ_x and σ_y are the respective standard deviations, and Wx, Wy are the wavelet power and abs symbol denote the absolute values from Wxy.

3. Comparison between observation (at Xisha station) and remote sensing for winds and SST

The AWS wind record is sub-sampled to match the QuikSCAT sampling time (in hours) near the station. Total 847 satellite / observation pairs are avaliable. The AWS wind observations is highly correlated to the QuikSCAT wind records (Fig. 1) with a linear correlation coefficient of 0.92 (Fig. 2a). The mean bias error (MBE) is examined to be -1.04 m/s, and root mean square error (RMSE) to be 1.30 m/s. The points are evenly distributed around the regression line. The result improves clearly if the outliers are taken out (Fig. 2b).

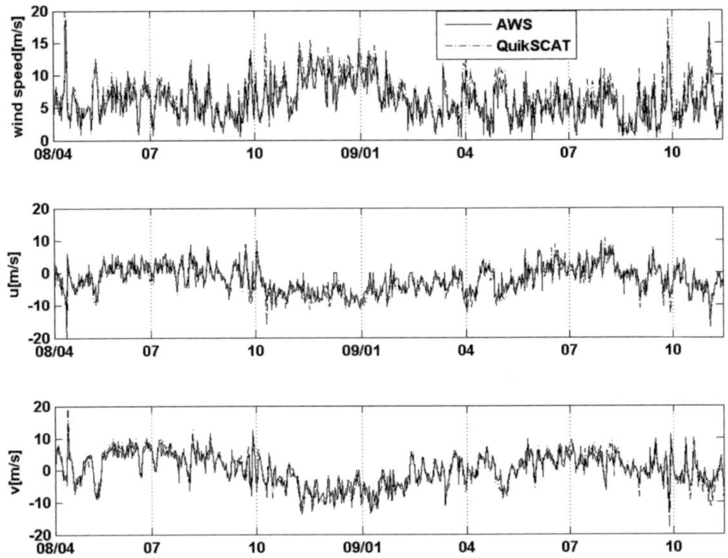

Figure 1. Comparison of QuikSCAT data and in-situ observation for the period April 2008-November 2009 at Xisha station: (a) wind speed, (b) zonal wind component, and (c) meridional wind component. The black line represents AWS observation, and the dashed line represents the QuikSCAT data.

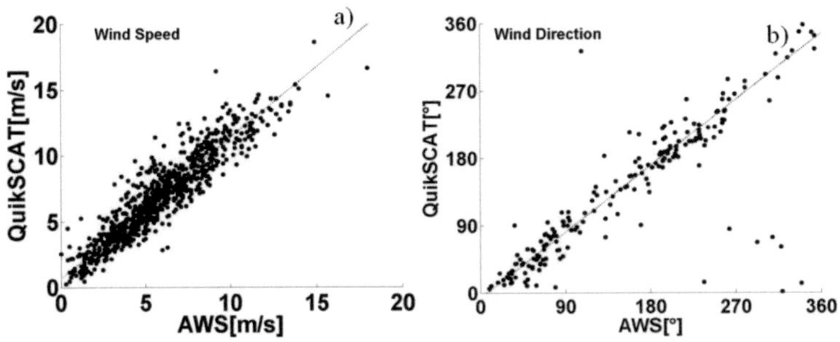

Figure 2. Scatter plot of QuikSCAT and AWS measurements in the period April 2008-November 2009: (a) wind speed linear regression (n=847), and (b) wind direction linear regression (n= 201). Dashed lines referred to the linear regression of the match-up pairs.

Additionally, the wind data (with 17 anti-phase outliers removed) is grouped into 16 groups (labeled as "N", "NNE", "NE", "NEE", "E", "SEE", "SE", "SSE", "S", "WSS", "WS", "WWS", "W", "WWN", "WN", "WNN") according to the direction, each group corresponding to an angle width of 22.5 degree, and the difference is examined individually for the 16 groups (Fig. 3a). The result shows that the wind speed data derived from the QuikSCAT is generally overestimated about 1.5m/s than the AWS data. The mean direction deviation is generally less than 20 degree.

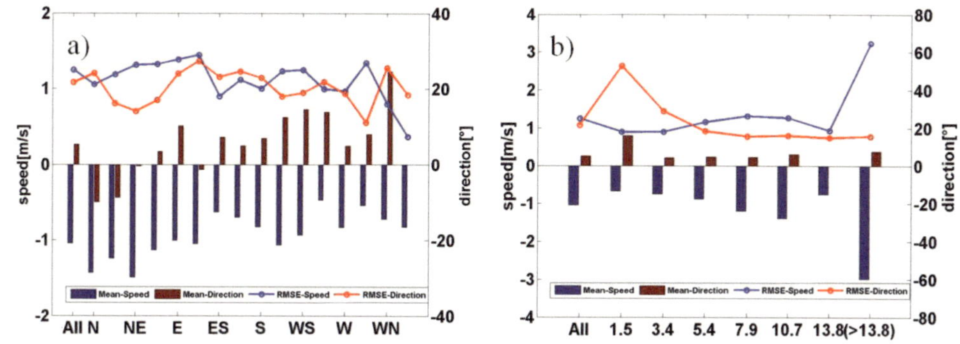

Figure 3. MBE (bars) and RMSE (lines) for speed and direction were classified as (a) a function of direction classes and as (b) a function of the Beaufort wind classes. The blue icons represent speed and red ones represent wind direction.

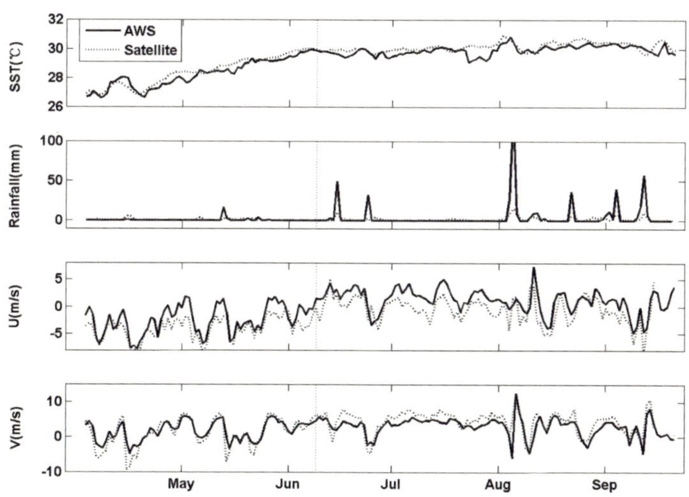

Figure 4. Comparisons of meteorological and oceanic variables in the period of March – September, 2010 at station Xisha: 5-day running mean of daily SST, rainfall, zonal wind and meridional wind. Black line represents in-situ observations and dashed line satellite data

Fig. 3b shows the MBE and the RMSE of wind speed and direction differences. It is apparent that RMSE (lines) becomes larger when the wind is strong or weak. The accuracy requirements of satellite wind speed estimates are satisfied if measurements values lower than 3 m/s are discarded [20]. Further, the consistency of the SST, rainfall and wind was checked between the remote sensing dataset and a buoy platform locates at Xisha from March 13, 2010 to August 31, 2010 (Fig. 4). The result shows that satellite-derived environmental variables are well consistent with observation. The difference between the SST and surface temperature exhibits significant seasonal variation, which can be mainly attributed to the seasonal typical stratification structure near the sea surface. The remote sensed SST can represent sea surface skin temperature, while in-situ SST measurements mainly characterize the temperature variability 2 m below the sea surface. In summer (winter), the skin temperature is much warmer (colder) than that below the sea surface due to the surface heating.

Fig.4a shows that SST increases and it reaches its peak before the summer monsoon onset. Then the local atmosphere slowly loses heat to cool down again after the summer monsoon onset. The satellite rainfall data also has few peaks in comparison with in-situ possibly because the satellite crossing region is deviation of Xisha station about 5-6 km (Fig. 4b). It has been clarified that year 2010 is a weak summer monsoon year (Fig. 4c and 4d). When the summer monsoon has a late onset, there is a shorter rainy season.

4. The variations of atmospheric variables recorded at Xisha station in the SCS during tropical cyclone passages

A total of 52 TCs passes through the SCS during 2008-2011, among which 21 TCs are found to be less than 400 km off the Xisha station (Fig. 5). The strongest TC occurred on September 23, 2008 with the maximum sustained wind speed in 125 knots at the northern XiSha. During TC passages, atmospheric variables, and air-sea condition will experience significant change, which also varies with the location of TC core. This study investigates atmospheric variables measured at Xisha station in the SCS when TCs passing by.

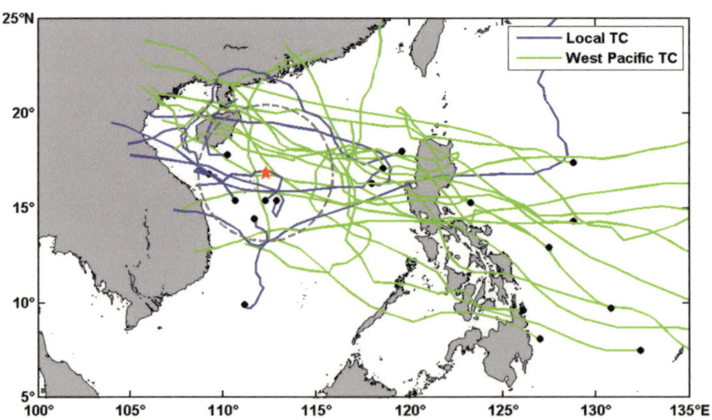

Figure 5. The track of 21 TCs passages within 400 km to the Xisha station during 2008-2011, black pots stand for the genesis location; red star means Xisha station, gray dashed line means the radius of 400 km off the Xisha.

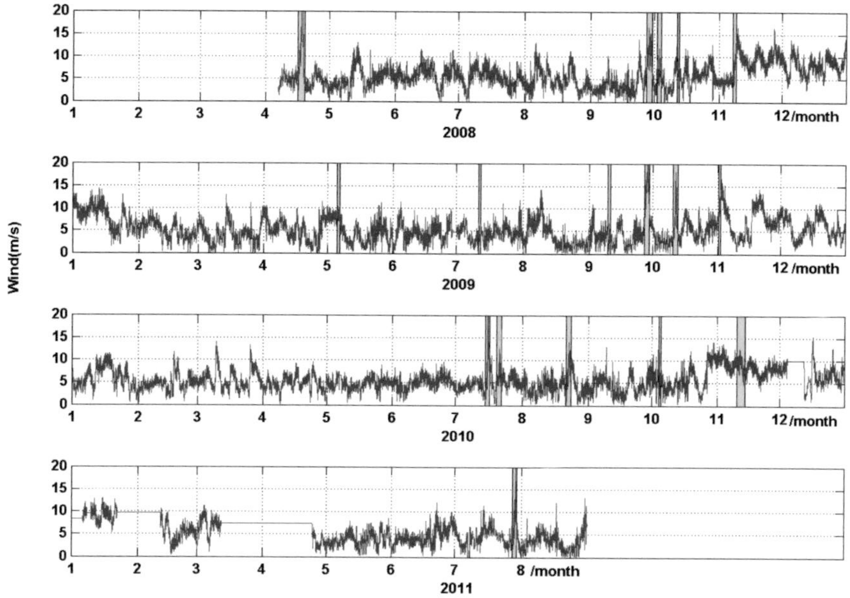

Figure 6. wind speed in AWS during April, 2008-October, 2011. Blue columns highlight the TC passages

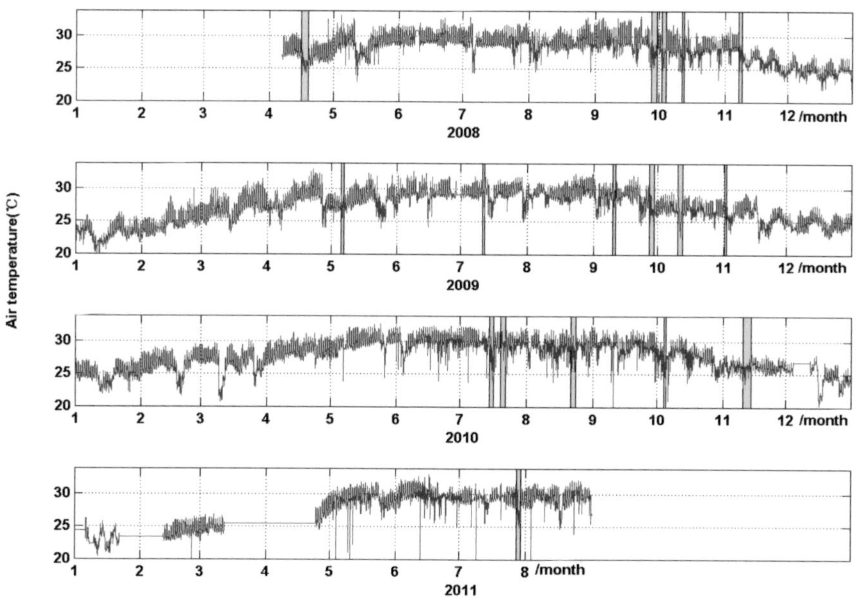

Figure 7. Air temperature in AWS during April, 2008-October, 2011. Blue columns highlight the TC passages.

While TC passes within less than 400km off the Xisha, abrupt increase in wind speed and decrease in air temperature can be clearly seen from the station records (Fig. 6 and 7). Wind speed dramatically increased over 10.8 m/s (according to Beaufort scale) and maximum can reach 32.5m/s on July 16, 2010. The Beaufort scale [21] is an empirical measure that relates wind speed to observed conditions at sea or on land. When wind speed reaches 10.8 m/s, it is described as Beaufort scale 6 and associated warning flag should be noticed. When TCs produce extremely powerful winds and torrential rain, they can introduce the strong surface heat exchange to maintain their development.TCs stir up water, leaving a cool wake behind them, which is caused by wind-driven mixing of cold water from deeper in the ocean and the warm surface waters (Eric, 2002). Cloud cover may also play a role in cooling the ocean, by shielding the ocean surface from direct sunlight before and slightly after the storm passage. All these effects can combine to produce a dramatic drop in the air temperature over a large area in just a few days.

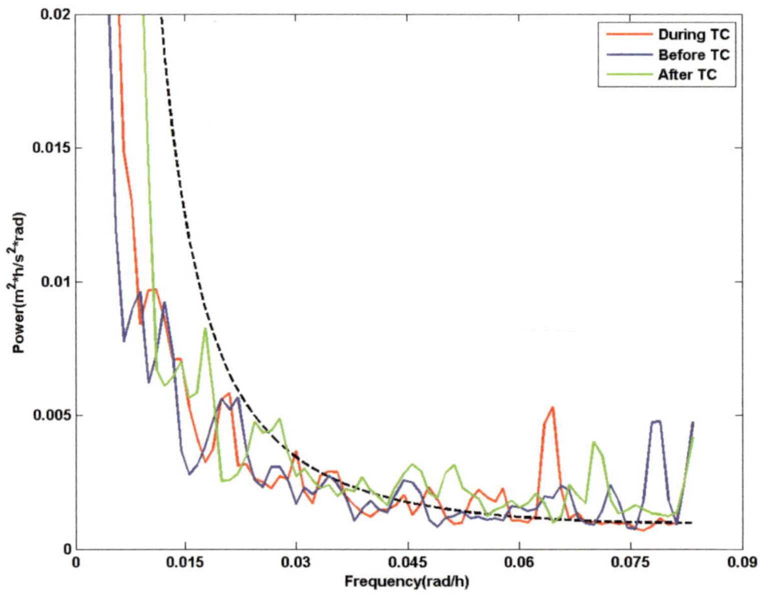

Figure 8. Wind velocity PSD for before, during, and after the TC passages. Black dashed line stands for 95% confidence interval.

Because of the TC strong intensity, they are also able to produce high-frequency waves in the local atmospheric environment. Therefore, spectral analysis is applied to the time-series of the wind velocity component for three periods; before, during and after the TC passage in order to study the discrepancy of velocity power spectrum (Fig. 8). The Power Spectrum Density (PSD) describes how the energy of a time series is distributed with frequency to estimate the spectral density of a random signal from a sequence of time samples with the fast fourier transform algorithm. The result presents that before the TC passes through the SCS, the wind velocity PSD shows semi-diurnal oscillation (12.85 hours). During the TC passages,

PSD is located at about 15.54 hours band, lower than that without TC passages. After TC passes, PSD gradually converted into the lower than semi-diurnal peak, occurred at about 14.06 hours band and its power decrease to some extent. The frequency change may be due to the TC physical processes After the TC passage, the power condition recovers to the normal state, which located between 12.85 hour and 15.54 hours band (about 14 hour).

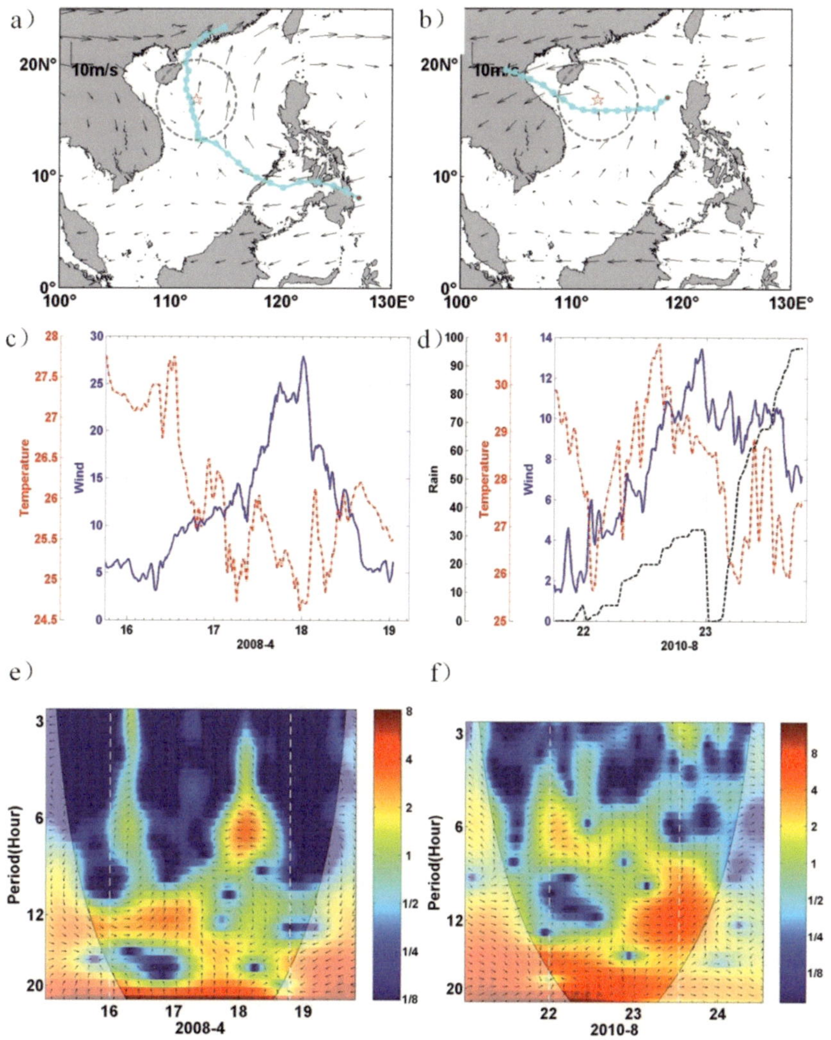

Figure 9. Case study of two TCs: a) Blue dashed line presents TC tack in April 16-18, 2008, dotted circle stands for the radius within 400km to Xisha station; black arrows show the steering flow; c) The time series of wind speed and air temperature observations at Xisha in April 16-18, 2008. e) The XWT of the standardized wind and air temperature in April 16-18, 2008. The 5% significance level against red noise is shown as a thick contour. The relative phase relationship is shown as arrows (with in-phase pointing right, anti-phase pointing left, and wind leading air temperature by 90 pointing straight down),and white dashed line means the period of this TC passage; b), d) and f) same as a), c) and (e) but for August 22-23, 2010.

In order to further understand the variations of atmospheric variables recorded at Xisha, two TCs cases are chosen in April 16-18, 2008 and August 22-23, 2010. The former formed in WNP, and crossed straightly through the Philippines into SCS, and then followed the northwards steering flow passing through Xisha region, making landfall in the South China (Fig. 9a). The latter generated in the eastern of SCS, then moving westward, and making landfall at the Vietnam coast (Fig. 9b). In the first case, the wind increased greatly at around 0400 UTC 16 April, 2008. When the AWS wind reaches the high peak, TC reached its maximum sustained wind speed in 95 knots at 111.4 E, 17.8N nearest Xisha station on about 0000UTC, 18 April, 2008. And at the same time, air temperature dropped from about 27.5 °C to 24.5 °C with the anti-phase wind (Fig. 9c). The other case is formed in northeastern of the SCS at about 1800 UTC August 20, 2010. The AWS wind started to increase greatly at around 2100 UTC 21 August, 2010. This TC sustained wind speed in 35 knots at 111.0 E, 16N on about 0000UTC, 23 August, 2010 when the AWS reached maximum. During this period, air temperature cooled from about 31.0 °C to 26.0°C and rain followed in just a few days (Fig 9d). Thus wavelet analysis is used to expose wind speed and air temperature common power and relative phase in time-frequency space (Fig. 9e and 9f). The air temperature and wind are significant common power in about 12-20 hour band during the TCs passage [22]. The XWT also shows that both of them are in anti phase with significant common power. This proved the previous view that a TC passage produces a dramatic drop in the air temperature over a large area in just a few days.

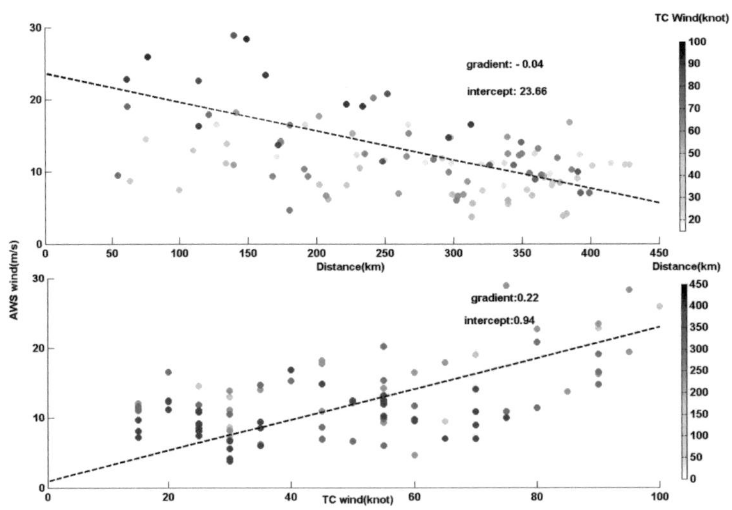

Figure 10. a) Scatter plot of TC distance to the Xisha station and AWS wind measurements. Dashed line represents linear fitting, the color bar illustrates the maximum wind speed at TC center (b) Scatter plot of the maximum wind speed at TC center and AWS wind measurements. Dashed line means linear fitting; the color bar stands for the distance between a TC to the Xisha station

TC tracks and intensity play an important role in the local atmospheric variations (Fig. 10a and 10b). The distance of a TC to the Xisha station is correlated to the local wind records

http://dx.doi.org/10.5772/50897

with a linear correlation coefficient of -0.57. The points are evenly distributed around the line. Stronger TC tends to induce larger wind variations at Xisha station (correlation coefficient=0.58).

5. Conclusions

Wind data at Xisha station during the period from April 2008 to November 2009 is compared with QuikSCAT and ASCAT. During the period, the crossing paths of both QuikSCAT and ASCAT are 5-6 km away from AWS Station, which is at 16º50′N, 112º20′E. The comparison confirms that QuikSCAT estimates of wind speed and direction are generally accurate, except for high wind speeds (> 13.8 m/s) and for the direction assessment under very low wind speeds (< 1.5 m/s).

There are 52 TCs through the SCS during 2008-2011, among which 21 TCs are found to be less than 400 km. When TCs passed through the Xisha, it induced the changes of wind field, air temperature and the occurrence of heavy rain. The wind velocity transforms the frequent from 12 hour to 15 hour during TC passage. The PSD become 14 hours after TCs passed through the Xisha. The wavelet analysis shows that the oscillations between air temperature and wind are common power and anti-phase. The variations of atmospheric variables at the Xisha station are larger if the TCs passing through are closer to the station or stronger in the intensity.

The statistical significance might be not very convincing due to such short observation period. Besides, current observation is only two dimensional, which is not enough to understand the vertical structure of the atmosphere.

Acknowledgements

This research is supported by the National Basic Research Program "973" program of China under contract Nos 2011CB403500 and 2011CB403504; the Chinese Academy of Sciences under contract No. KZCX2-YW-Y202; the National Natural Science Foundation of China under contract No. U0733002.

Author details

Dongxiao Wang*, Jian Li, Lei Yang and Yunkai He

*Address all correspondence to: pearlriverbay@gmail.com

State Key Laboratory of Tropical Oceanography, South China Sea Institute of Oceanology, Chinese Academy of Sciences, Guangzhou,, China

References

[1] Chan, J. C. L. (2007). Interannual variations of intense typhoon activity. Tellus Printed in Singapore., 59A, 455-460.

[2] Goh, A. Z. C., & Chan, J. C. L. (2010). Interannual and interdecadal variations of tropical cyclone activity in the South China Sea. *Int. J. Climatol.*, 30, 827-843.

[3] Wang, L., Lau, K. H., Fung, C. H., & Gan, J. P. (2007). The relative vorticity of ocean surface winds from the QuikSCAT satellite and its effects on the geneses of tropical cyclones in the South China Sea. *Tellus*, 59A, 562-569.

[4] Camargo, S. J., Robertson, A. W., Gaffney, S. J., Smyth, P., & Ghil, M. (2007). Cluster analysis of typhoon tracks. Part I: General properties. *J. Climate*, 20, 3635-3653.

[5] Holland, G. J. (1995). Scale interaction in the western Pacific monsoon. *Meteor. Atmos. Sci.*, 56, 57-79.

[6] Dvorak, V. F. (1975). Tropical cyclone intensity analysis and forecasting from satellite imagery. *Mon. Wea. Rev.*, 103, 420-464.

[7] Webster, P. J., Holland, G. J., Curry, J. A., & Chang, H. R. (2005). Changes in tropical cyclone number, duration, and intensity, in warming environment. *Science*, 309, 1844-1846.

[8] Knaff, J. A., Demaria, M., Molenar, A., Sampson, C. R., & Seybold, M. G. (2011). An Automated, Objective, Multiple-Satellite-Platform Tropical Cyclone Surface Wind Analysis. *Journal of applied meteorology and climatology*, 50, 2149-2166.

[9] Lau, K. H., Zhang, Z., Lam, H., et al. (2003). Numerical simulation of a South China Sea typhoon Leo (1999). *Meteoro Atmos Phys*, 83, 147-161.

[10] Li, J., Wang, A., Hou, E., et al. (2004). A numerical prediction experiment of track and heavy rainfall about Typhoon Fitow. J Tropical Oceanog (In chinese)., 23(1), 7-10.

[11] Zhao, Y., Wang, B., Liang, X., et al. (2005). Improved track forecasting of a typhoon reaching landfall from four-dimensional variational data assimilation of AMSU-A retrieved data. *J Geophys Res*, 110, D14101.

[12] Dell'osso, L., & Bengtsson, L. (1985). Prediction of a typhoon using a fine-mesh NWP model. *Tellus A*, 37A, 97-105, doi: 10.1111/j.1600-0870.1985.tb00273.x.

[13] Yang, L., Wang, D., & Peng, S. (2012). Comparison between MM5 simulations and satellite measurements during Typhoon Chanchu (2006) in the South China Sea. *Acta Oceanologica Sinica*, 31(2), 33-44, DOI: 10.1007/s13131-012-0190-3.

[14] Kubota, J., & Chan, J. C. L. (2009). Interdecadal variability of tropical cyclone landfall in the Philippines from 1902 to 2005. *Geophys. Res. Lett.*, 36, L12802.

[15] Jen C. H. (2010). Quaternary environment change in Taiwan: the typhoon events observation by weather data in mountainous region. EGU General Assembly. 2010, held 2-7 May, 2010 in Vienna, Austria , 8779.

[16] Li, J., Wang, D., Chen, J., & Y., L. (2012). Comparison of Remote Sensing Data and insitu Observation for winds during the development of the South China Sea Monsoon. Chinese Journal of Oceanology and Limnology.

[17] Satheesan, K., Sarkar, A., Parekh, A., et al. (2007). Comparison of wind data from QuikSCAT and buoys in the Indian Ocean. *Int. J. Remote Sensing*, 28(10), 2375-2382.

[18] Kalnay, E., Kanamitsu, M., Kistler, R., et al. (1996). The NCEP/NCAR 40-year reanalysis project. *Bull. Amer. Meteoro. Soc.*, 77, 437-470.

[19] Jevrejeva, S. (2003). Influence of the Arctic Oscillation and El Nin˜o-Southern Oscillation (ENSO) on ice conditions in the Baltic Sea: The wavelet approach. *J. Geophys. Res.*, 108(D21), 4677, doi: 10.1029/2003JD003417.

[20] Pensieri, S., Bozzano, R., & Schiano, M. E. (2010). Comparison between QuikSCAT and buoy wind data in the Ligurian Sea. *Journal of Marine Systems*, 81(4), 286-296.

[21] Huler, S. (2004). Defining the Wind: The Beaufort Scale, and How a 19th-Century Admiral Turned Science into Poetry. *Crown*, 1-40004-884-2.

[22] Grinsted, A., Moore, J. C., & Jevrejeva, S. (2004). Application of the cross wavelet transform and wavelet coherence to geophysical time series. *Nonlinear Processes in Geophysics*, 11, 561-566.

Chapter 5

Characteristics of Hurricane Ike During Its Passage over Houston, Texas

Gunnar W. Schade

Additional information is available at the end of the chapter

http://dx.doi.org/10.5772/51344

1. Introduction

Hurricane impacts, real and perceived, are most obvious when a hurricane makes landfall in populated areas. While hurricane track and strength forecasting has clearly improved in the last decade, increasing coastal development puts increasing population numbers in harm's way [1]. In addition, the modified coastal environment with its human infrastructure is affected by coastal flooding from the storm's surge, inland flooding from torrential rains, and damaging hurricane winds. Recent hurricane research has sought to characterize land-falling hurricanes better [2-7], particularly in order to understand the dynamics and development of hurricane winds during landfall and the associated potential for wind damage to structures and the natural environment. The necessary safety prerequisites to instrumentation observing land-falling hurricanes, and the amount of data they collect demand a high organization, quick response to meteorological developments, careful data analysis, and patience. Thus, as this research is only approximately a decade old few reports have appeared in the peer-reviewed literature so far.

Meteorological measurements inside urban areas are also relatively rare as guidelines for proper setup of representative weather stations usually lead to conflicts in such areas. However, meteorological measurements, as well as weather and impact forecasts in and for conurbations are becoming more important as urban populations increase. An increased (meteorological) focus has been put on the urban environment [8-12], and guidelines have been developed for urban measurements and site qualifications [13-15]. Micrometeorological observations in urban areas are especially challenging but many more observational data sets have been acquired in the last decade since Roth's seminal review in 2000 [9, 16]. One of these is being acquired in Houston, Texas, since summer 2007 as part of a project to study atmospheric turbulence, and anthropogenic and biogenic trace gas fluxes over a typical urban landscape

[17, 18]. Approximately a year after project measurements had begun, Hurricane Ike approached the Gulf Coast with a forecast track right through the City of Houston. While the author's installations on the communications' tower used for the study were designed to withstand high winds, the tower itself was not rated for a category three hurricane, and there was thus a good chance it would topple if Ike maintained its strength upon landfall. On September 12/13, 2008, at landfall near Galveston, Texas, Ike was finally rated "only" a category two storm on the Saffir-Simpson scale. The tower survived. However, due to the impacts of Ike's enormous size and wind field the storm caused widespread disaster and power loss to approximately 2.1 million customers in the Houston metro area [19], and is now considered the USA's second most costly hurricane in history at nearly 30 billion USD of estimated, accumulated damages [20].

A major impact of Ike to people and infrastructure in the Houston metro area stemmed from debris, both from natural vegetation and damaged human infrastructure [21]. Particularly the amount of damage to the energy (electricity) infrastructure was staggering, leaving sections of the metro area population without power for more than three weeks after the storm's passage [19]. What was highly visible in Houston long before the storm, and had been concluded by us as the major culprit [22], was the lack of proper tree trimming leading to tree-debris-caused power line failures. Within the first year after the storm, it had been concluded that this was indeed the leading cause for the amount of failures, and that in conclusion future activities would focus on appropriate tree trimming [23, 24].

As the instrumented communication's tower north of downtown Houston rode out the storm, albeit partially damaged, and power to the site was never lost due to a local diesel generator, it was possible to record meteorological and air quality data both during the storm and its aftermath. This chapter will focus on the storm's meteorological characteristics as measured over the urban area north of downtown Houston. They represent, so far, the only recorded hurricane passage over an urban environment [22]. Firstly, there will be a review of the installation's salient features and important past observations, followed by a brief description the hurricane's development. The following sections will focus on the development of Ike's winds over Houston in comparison to past findings, and their impacts in the neighborhood of the site. Lastly, connections between the hurricane's thermodynamics, its rainbands and winds, and air quality will be explored. A more detailed view of the latter shall be reported elsewhere.

2. The Yellow Cab tower site

The Greater Houston Transportation Co. owns and operates a 91 m tall communications tower at 1406 Hays Street, 77009 Houston, TX, approximately 3.5 km NNE of downtown Houston (29.789° N, 95.354° W, 14 m above sea level). In 2007 the tower was equipped with meteorological and micrometeorological instrumentation to measure standard meteorological parameters plus atmospheric turbulence and surface energy exchanges. In June 2008, the existing 4-level gradient measurements were expanded by a level at 49 m above ground level (agl), and an extra sonic anemometer at 40 m agl. Details of the instrumentation can be found in [17], downloadable from the project's website. On the top level, a MetOne model 034B wind speed (ws) and direction (wd) sensor was used, and on the lower levels MetOne 014A ws only models with standard metal cups were used. Both sensors have a starting threshold of <0.5 m s^{-1} and distance constants of <= 4.5 m, equivalent to a 95% of maximum response time of approximately 3 seconds in a typical 9 m s^{-1} wind. Data were recorded in 10-s intervals and stored as 1-min averages and standard deviations except atmospheric pressure, which is only recorded every 15 minutes, but was smoothed for the analysis purposes of this study. Micrometeorological data from the installed sonic anemometer at 60 m agl was recorded at 10 Hz up to approximately midnight on 12 September 2008 when increasing rainfall began to completely cut the signal. The sonic anemometer then did not recover until two days after the storm. Rainfall was recorded with a tipping bucket (model TE525) on the E side of the tower, ten meters agl, and atmospheric pressure from within the data logger enclosure at two meters agl. Lastly, carbon monoxide was measured by a model 48CTL gas filter correlation instrument, corrected for background drift every two hours and calibrated once a week (precision ±20 ppb). The instrument samples air from inlets at 40 and 60 m agl on the tower.

Figure 1 shows a picture of the instrumented tower in summer 2008 (a), and a bird's eye view of the surrounding area with labels (b). A more detailed description of the underlying urban surface is given in [25] and our TARC report [17]. Of particular interest with respect to wind measurements is the location of the sensors relative to the lattice tower structure itself. Figure 2 shows the setup on the top level (60 m agl) and an example of the lower level installations.

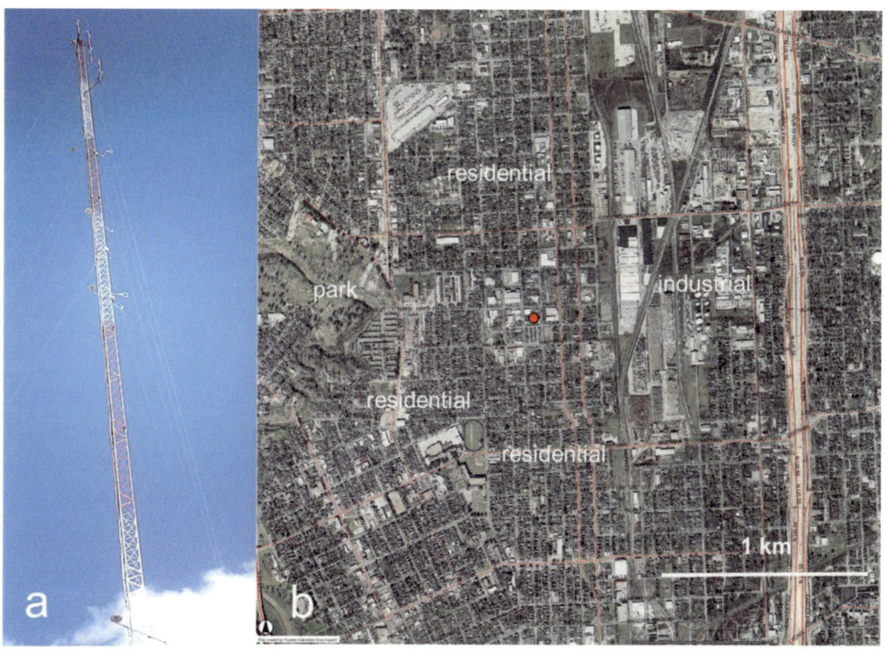

Figure 1. a) Picture of the top four meteorological installations on the flux tower during summer 2008. Note temporary installation of an additional sonic anemometer at the 40 m agl. (b) Annotated bird's eye view of the tower location (red dot) north of downtown Houston. Red lines indicate major thoroughfares.

As the wind climatology for Houston indicated rare occasions of northerly winds, the sonic and the lower level cup anemometers were installed on the south side of the tower. The top level ws/wd sensor was installed as a "backup" sensor on the NW side of the tower. As shown in Figures 1&2, the triangular tower lattice structure has a side length of 60 cm (2 ft), thus is bound to affect wind flows towards the cup anemometers for average wind directions between approximately 320 and 10 degrees as the sensors are within three times the largest horizontal tower dimension. A similar argument holds for the sonic anemometer at similar wind directions, and for the ws/wd sensor for SE wind directions. Though the effects may be smaller in both the latter cases due to larger distances from the tower, there is an additional electronics box installed on the tower at that level, which can cause an extra wake.

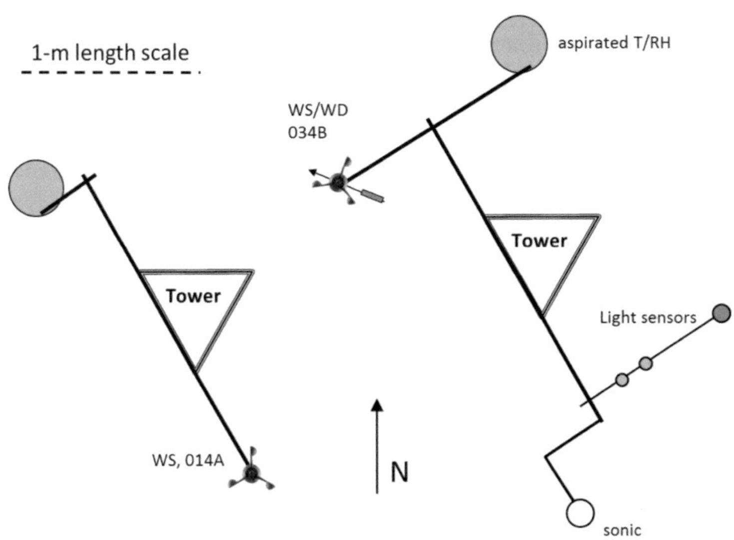

Figure 2. Meteorological instrumentation installed at the Yellow Cab tower in Houston at 60 m agl (right), and the lower levels (49, 40, 20, 13 m agl; left). The tower has a side length of 2 ft with vertical bars of 1 ½ inches diameter. Installed horizontal beams (solid black lines) consist mostly of aluminum pipes (Campbell Scientific Inc., Logan, Utah, USA). Note that the aspirated radiation shields (MetOne model 075B) are installed with an off-center beam.

Measurements from the tower since late summer 2007 were used to determine the structure of atmospheric turbulence at this urban site, and the underlying surfaces' roughness lengths (z_0) and displacement heights (d). Figure 3 shows the observed behavior of normalized longitudinal and vertical winds – as relevant for this study – with the atmospheric stability parameter ζ=(z-d)/L for daytime unstable conditions, where is z is measurement height and L is the Obhukov length. We observed surprisingly low directional heterogeneity in the data, thus these results were pooled. A comparison to the review of Roth [16] listed in Fig. 3 showed few differences as compared to past findings over urban areas.

A combination of methods to determine z_0 and d [26] was used to arrive at values of $z_0 = 1\pm0.1$ m, and $d = 8^{+3}_{-2}$ m [17]. The distribution of values together with an analysis of building and tree heights in the area revealed that d is dominated by the tree canopy in this area [17]; details will be published elsewhere. Some heterogeneity was observed for heat and trace gas fluxes, but details shall also be given elsewhere. Here, the surface heterogeneity in the immediate surroundings of the tower is relevant: Within a radius of 150 m, the surface is largely impervious and dominated by warehouse-style buildings with different roof angles and heights. The tallest building is a large, approximately 50×70 m^2 structure to the NW that can be easily spotted in Figure 1b. There are very few tall trees within a 100 m radius from the tower.

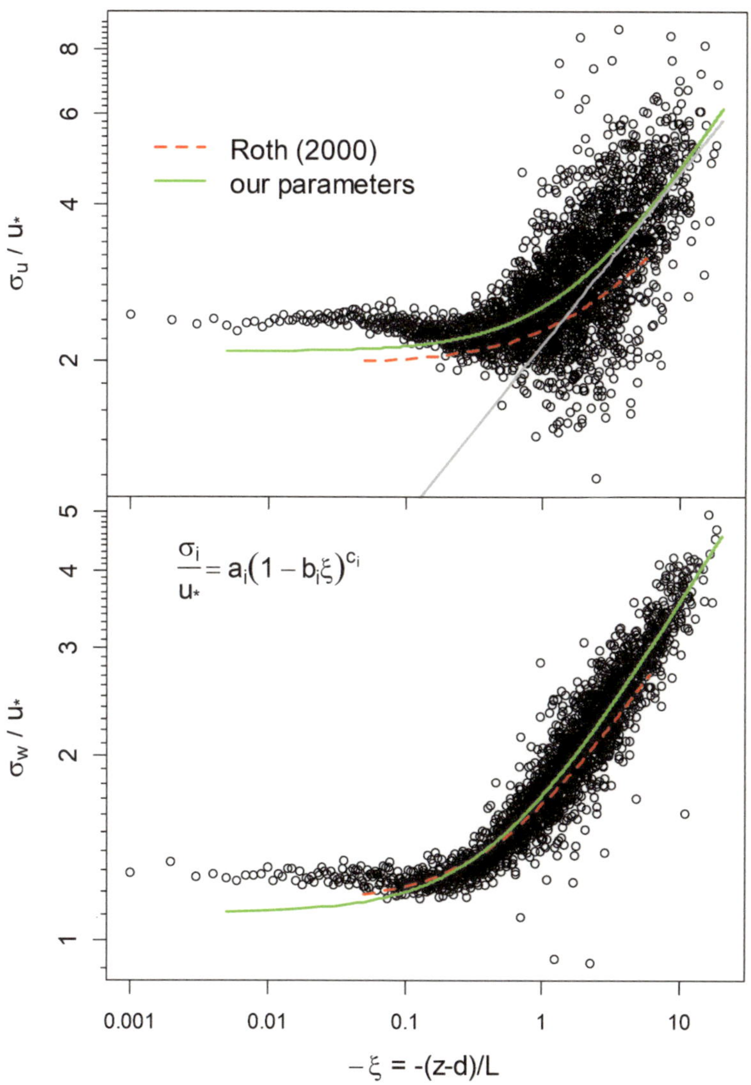

Figure 3. Normalized longitudinal and vertical wind standard deviations as function of atmospheric stability for unstable conditions, compared between our site and other urban data as summarized by Roth [16]. The grey line shows the expected c_i=⅓ slope. For all i=u,v,w, we observed neutral stability values close to the literature data, namely 2.4, 1.8, and 1.3.

Despite some obvious surface heterogeneity, the lack of significant directional differences in the measured turbulence can be explained by the z=60 m agl measurement height, which is approximately six times the displacement height and therefore well above the recommended height for such measurements [16], inside the inertial sublayer. This height was initially chosen to integrate our flux measurements over a larger, more representative urban surface area. Flux footprint estimates [25] indicate that under typical wind speeds (4-6 m s^{-1}) and turbulence conditions (friction velocity, u_*=0.4-0.6 m s^{-1}) at this site, 90% of measured fluxes are expected to come from less than 1.5 km distances, with maximum impacts from locations <=500 m from the tower [25], and total footprint sizes of 2-4 km^2. However, for wind speeds between 20-30 m s^{-1} and associated turbulence discussed below, maximum impact areas move significantly closer to the tower (<=350 m) and footprint size contracts, so that a larger effect of the surface and its heterogeneity close to the tower on the measurements is to be expected. This is also evident through a decrease of the displacement height with observed wind speeds (or friction velocities) [17], likely because of the lack of tall trees closer to the tower.

3. Hurricane Ike

Ike was a somewhat uncharacteristic hurricane due to its very large eye and wind field, filling nearly the entire Gulf of Mexico on 12 September 2008. Though its wind speeds classified Ike as a category two hurricane at landfall (maximum winds between 96 and 110 mph = 43-49 m s^{-1}), a more holistic view called the *hurricane severity index* [27] ranks it similar to such disastrous landfalling hurricanes as Andrew in 1992 and Ivan in 2004. Most of Hurricane Ike's history and measures, including its track, precipitation, and measured or estimated winds are described online [28], and shall not be repeated here. A track map through Houston and a brief analysis of our meteorological data were also given earlier [22]. For comparison purposes to this study, Figure 4 shows two important meteorological variables: Fig. 4a shows a precipitation radar image (KHGX Houston, located at the green-blue dot south of the eye) from the period of maximum wind speeds, approximately one hour before the minimum locally measured pressure at the tower site; Fig. 4b shows gridded, calculated maximum sustained wind speeds for Harris County, assembled from the original GIS data [28], which have generally been shown on a much larger scale [27, 29]. Note that Ike preserved its very large eye for a comparatively long time after landfall, and that the City of Houston, roughly outlined by the *Beltway*, was well within Ike's eyewall. As a result of this, several additional rain bands passing over, and the flat Texas coastal terrain, widespread flooding occurred in central and north Houston. Particularly, the *White Oak Bayou* watershed area, at the SE corner of which the tower site is located, was listed as "Major Flooding / House flooding area from rainfall". The same area was calculated to have experienced category one (74-95 mph = 33-43 m s^{-1}) hurricane wind speeds, with a maximum of 38 m s^{-1} estimated for the tower site and much of central Houston (Fig. 4b). However, it needs to be pointed out here that these are NOAA AOML Hurricane Research Division (HRD) *surface wind* data that are – in this case – valid only for "open terrain exposure over land". They are a standardized product reflecting speeds at 10 m

http://dx.doi.org/10.5772/51344

agl for a flat surface (d=0) without significant roughness elements (z_0<=0.1 m). Actually measured speeds are discussed below.

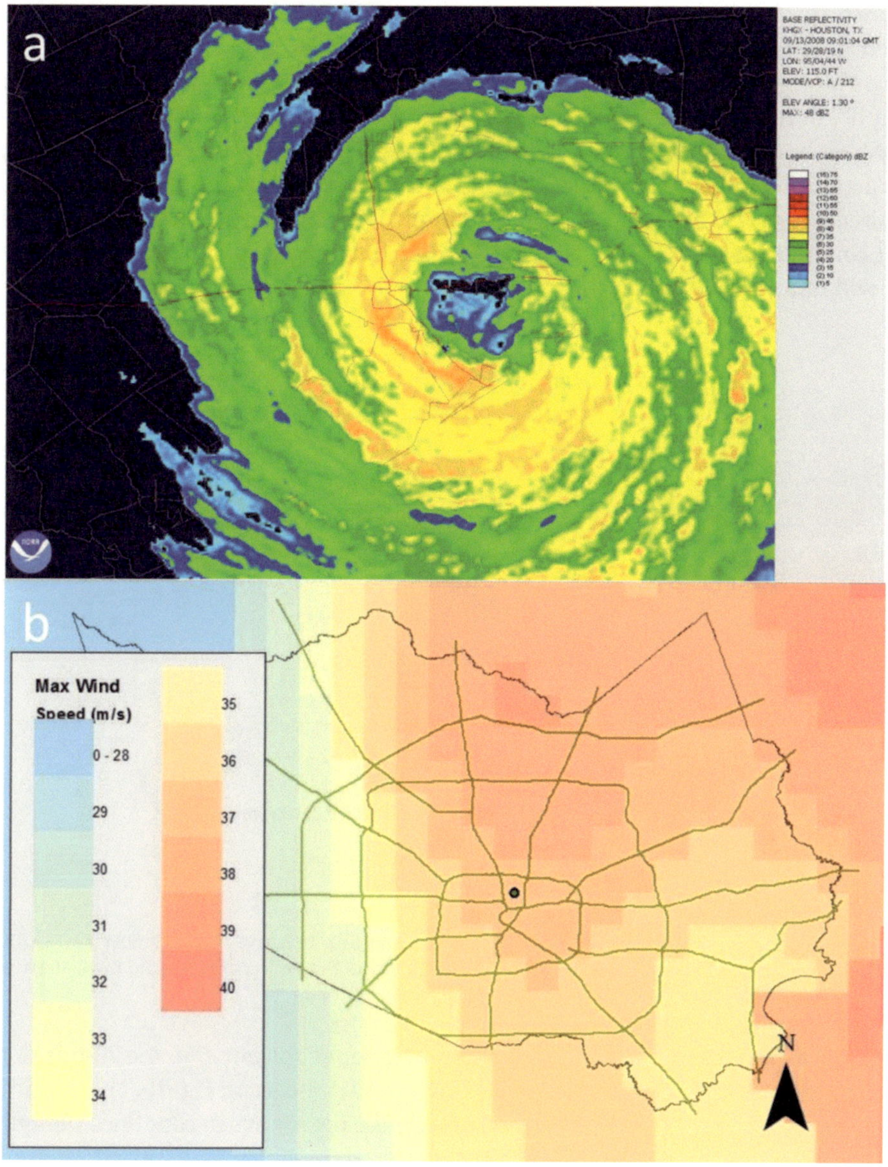

Figure 4. a) Precipitation radar image from 13 September, 09:00 GMT (03:00 CST) overlaid on a Texas county map (grey lines) and including interstate highways (red lines), the circular one in the center depicting Houston's *Beltway*, which extends for approximately 19 km in its E-W direction. (b) Maximum sustained wind speeds (NOAA Atlantic Oceanographic & Meteorological Laboratory Hurricane Research Division H*Wind product) during Ike in Harris County, Texas, overlaid on the county border (black line) and major thoroughfares (green lines). The central dot marks the tower location.

4. Wind measurements

4.1. Wind speeds and turbulence intensities

In Figure 5 are depicted measured wind speeds at 60 and 20 m agl from the Yellow Cab tower for 12th (DOY 256) and 13th (DOY 257) September 2008. Wind direction is included in the form of arrows along the bottom, and approximate timing of rain band passages along the top of the graph. For northerly winds, which occurred in the afternoon of DOY 256 and for approximately four hours around midnight at the height of the storm, the measured 20-m agl wind speeds are clearly impacted by the tower structure as explained above. As the eye passed and wind direction shifted to the NW, the sensor moved out of the tower's wake, and recorded its highest 1-min values of 22-25 m s^{-1} around 4 am Central Standard Time (CST = CDT - 1 h). This timing will be addressed again below.

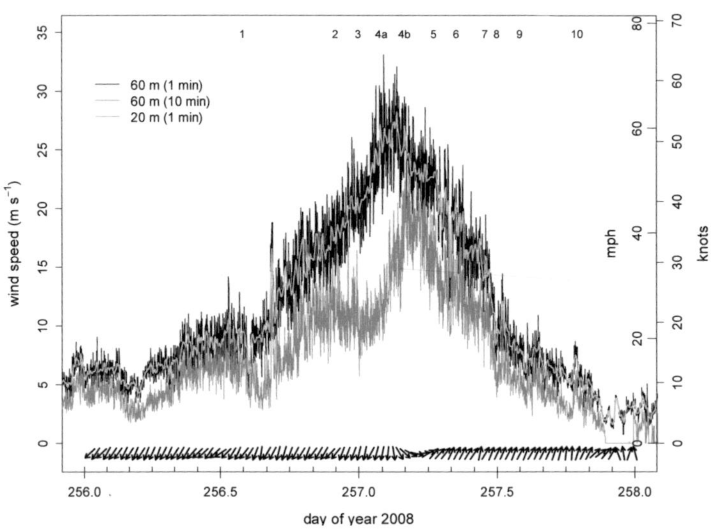

Figure 5. Measured wind speeds during Hurricane Ike's passage over Houston on September 12-13, 2008. Arrows along the bottom depict wind direction, numbers on the top locate the timing of rain band passages with labels 4a and 4b identifying the eyewall.

Considering a displacement height of close to ten meters for our site, the 20-m agl measurements come closest to the standardized H*Wind product shown in Fig. 4b. The large difference between the two (25 vs. 38 m s^{-1}) is associated with the higher urban roughness length. Higher roughness lengths increase surface friction and therefore reduce wind speed.

Fig. 5 includes 10-min average wind speeds for the top level sensor and therefore allows a direct comparison between sustained winds and 1-min gustiness (10-s mean winds were not recorded). Only the latter revealed hurricane-force winds (>33 m s^{-1}) at the top level, 60 m agl, while sustained (10-min) tropical storm force winds (>17 m s^{-1}) were recorded for nearly 12 hours. However, extrapolating sustained top level to the 20-m agl wind speeds using the log-wind-law

$$\hat{U} = u*/k \times \ln((z\text{-}d)/z_0) \tag{1}$$

where $\hat{U}$ is 10-min average wind speed, u_* is friction velocity, and k=0.4 is von Karmann's constant, reduces the time of 17+ m s^{-1} wind speeds to a mere 15 min. This analyses shows that the rough urban surface was very efficient in slowing down Ike's winds. Other surface maximum wind speed observations in the Houston urban area, compiled by Berg [20], ranged from 16 (west Houston) to 65 (Houston Hobby airport, south Houston) knots (see Fig. 5 for speed conversions), confirming that hurricane force winds were rare in the urban canopy even in the eyewall.

Higher surface roughness in the urban area though increases atmospheric turbulence. The latter was analyzed in two ways, investigating drag or wind force, and turbulence intensities, I_U. The former was calculated following [30], also cited in [31], from the 20-m agl sustained wind speed data in combination with air density. The latter was calculated from 15-min non-overlapping segments for both the sonic and cup anemometers, except during the last six hours before midnight when lower data density due to rain required half-hourly or hourly averages. In Figure 6 we plotted the resulting wind force per area and its standard deviation. Mean values characterize *static*, while fluctuations characterize *dynamic* wind load on surface roughness elements. Under "normal" conditions (DOY 256 morning), loads rarely exceed 1 kg m^{-1} s^{-2}, but during the hurricane, loads increased by a factor of 5-10 in the eyewall.

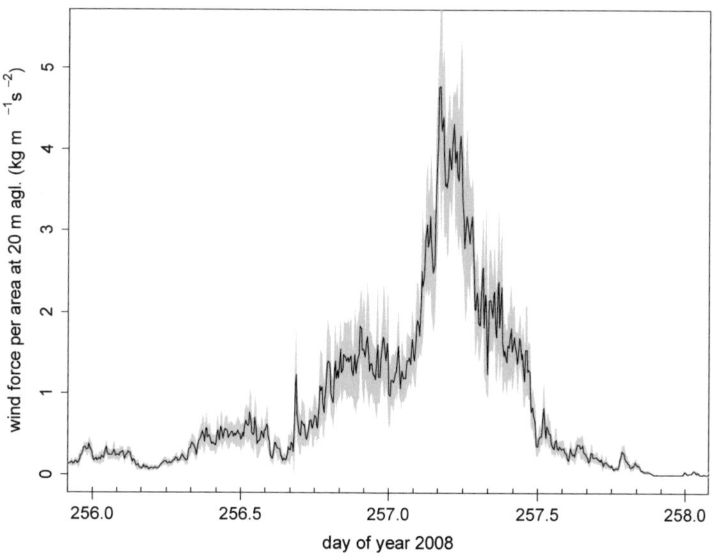

Figure 6. Wind force, proportional to $\hat{U}^2$, time series during Hurricane Ike's passage over Houston. The grey swath is an estimate of its standard deviation.

In Figure 7 are plotted turbulence intensities time series as calculated from the sonic anemometer data and the top level wind sensor, alongside the storm's pressure, wind direction, and rain band timing characteristics. Due to the higher data density of 10 Hz vs. 0.017 Hz, values are much larger for the sonic vs. the cup anemometer. Prior to the storm, I_U is influenced by atmospheric stability, leading to higher values during daytime unstable conditions. Values further increased when wind direction was northerly due to a reduction in wind speed but not its standard deviation in the tower's wake. Sonic-derived longitudinal (I_u) and lateral (I_v) turbulence intensities can be compared to the high frequency data obtained from anemometers installed near the coast line during the landfall of other hurricanes [3-6, 32].

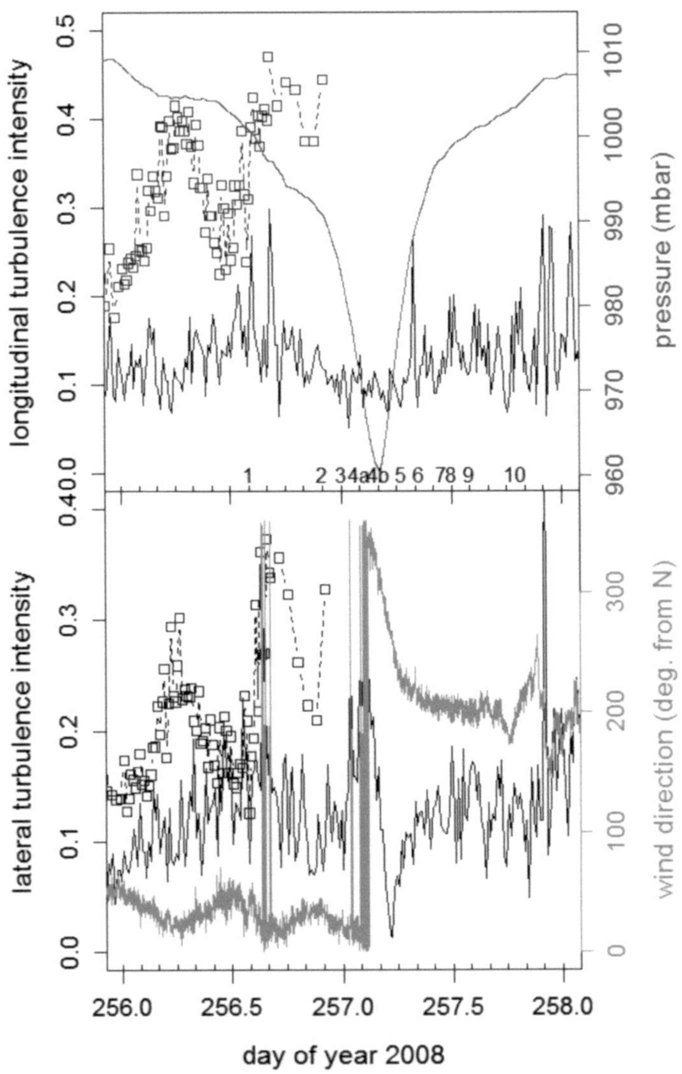

Figure 7. Turbulence intensities as measured by the sonic and top level cup anemometers as compared to storm passage characteristics of pressure (top panel) and wind direction (bottom panel). Lateral turbulence intensity of the 034B sensor depicted in the bottom panel is assessed from the 1-min wind direction fluctuation (right axis) and plotted as $10 \times \sigma_{wd}/360$ (left axis). Note the surprise lull under northwesterly wind directions towards the end of the eyewall passage.

As expected, values are, in general, larger than those obtained over the relatively smooth surfaces encountered in those previous studies. However, the expected dependence of I_U on surface roughness length is clear also in previous data [6]. In addition, I_U also depends on measurement height and wind speed. Turbulence intensity data obtained from a tall tower over urban Beijing [33] are roughly consistent with the above results, I_u = 0.3-0.4, and I_v=0.25-0.35 at similar measurement heights. However, while a mean I_v/I_u ratio of 0.87 was obtained, other studies such as [33]and [6] obtained 0.75, close to the expected value from MO theory, suggesting that the tower's wake influences reduced I_u more strongly than I_v. Nevertheless, use of the predictive formulas listed in [33] to compare the results to resulted in a reasonable fit using the formula

$$I_u = 0.765 \times \left(10/z\right)^{0.447} \tag{2}$$

which leads to a value of 0.34 compared to a mean measured value of 0.29±0.09 (1 sd), not including wake-affected values for the two days entering the calculation (DOYs 255&256).

Wind speed influences on I_U can be observed from the top level cup anemometer data. As the storm approached, I_U dropped slightly and became less variable, more or less consistent with the pressure development. A similar result was presented in [33] for winds measured over Beijing, while not as clear for the coastal hurricane land fall measurements [4, 6].

A closer inspection of observed wind speeds and associated turbulence intensities in Figures 5 and 7 shows that rain band passage affected both measures. Wind speeds increased just prior to or shortly after rain band passage. Spikes are visible both prior to and after the first outer rain band passed over Houston, as well as prior to nearly all outer rain band passages. A similar argument can be made for the eye wall passage itself: Label 4a characterizes passage of the NW side of the eye wall, while label 4b characterizes passage of the SW side of the eye wall. Both were associated with heavy rainfall, but the period in between experienced slightly lower rainfall rates. In all these cases, short-term increased wind speeds were likely associated with mesoscale circulations within the storm associated with descending air between rain bands accelerating towards an inner rain band. Convection inside the rain band causes heavy rainfall and was generally associated with lulls in wind speed. In contrast, as can be seen from Figure 7 it was often associated with short-term increases of turbulence intensity as a likely effect of stronger convective activity.

4.2. Mesoscale storm characteristics

With increasing wind speed in storms, large scale wave structures are expected to form, often called *rolls*. These have been investigated for hurricanes before through Doppler radar, integral length scales (L_x, x=u,v,w), and wind speed spectral analyses [2, 5-7, 34]. Several of these authors found that both sub-kilometer and meso-gamma scale (several kilometers) features can be observed in hurricanes. Compared to observations at lower wind speeds, these features lead to higher values at lower frequencies in the (normalized) wind power spectrum, particularly between 0.01 and 0.001 reduced frequency values. Some of these analyses have been repeated here; however, the loss of sonic anemometer data during rain prevents a detailed spectral analysis, and the data acquisition frequency of the cup anemometers limits integral length scale and spectral analysis.

R software [35] was used to calculate normalized, partially smoothed spectra from 2-hour de-trended segments of both sonic and cup anemometer data. Integral length scales were calculated from the sonic data by (i) selecting non-overlapping 10-min segments and calculating the auto-correlation function (acf), (ii) fitting an exponential decay curve to the acf, and (iii) integrating the exponential function out to twice the first zero-crossing observed in the acf. The procedure was similar to that described in [6] and [4].

Here, pre-storm wind speed spectra from both the top level sonic and cup anemometers were compared to height-of-storm cup anemometer data. A short-coming of the previous works is the lack of comparison between observed hurricane wind spectra and "normal" wind speed spectra at the same locations. Figure 8 shows the calculated power spectra for the 2-h, neutral stability longitudinal wind segments.

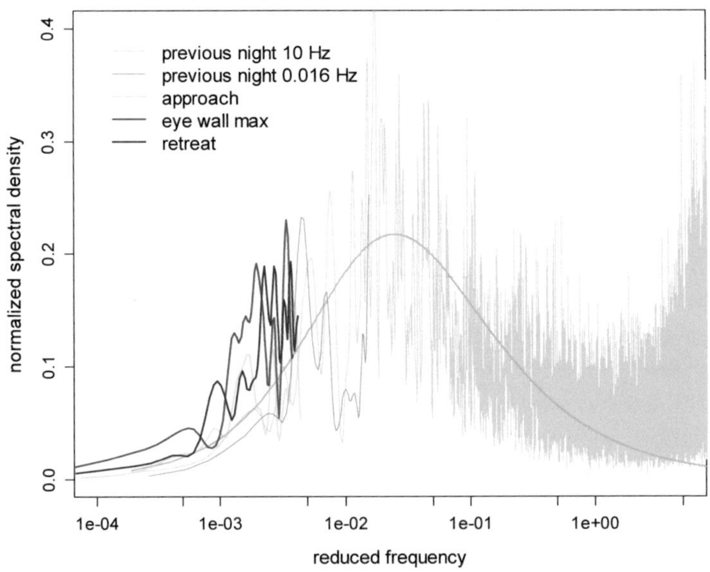

Figure 8. A comparison of longitudinal wind speed spectra between the night before the hurricane (DOY 256, hours 01:00-03:00 CST) and three periods of the highest encountered wind speeds before, during, and after the eyewall maximum. Smoothing was limited to highlight differences. The thick, dark grey line represents a standard spectrum. Note that the grey spectra from the previous night were obtained at an average wind speed of 6 m s^{-1}, the offset being due to a slightly higher value recorded by the cup anemometer. Wind speeds for the blue spectra were all above 22 m s^{-1} (Fig. 3). To match power density levels (y-axis) the blue spectra were additionally normalized to the wind speeds' standard deviation ratios.

The grey spectra match surprisingly well between the top level sensors at lower frequencies. At the same time, they are well characterized by the standard, smoothed "perturbed terrain" model spectrum [4], which shows a maximum at 2-3×10^{-2} and falls off below

1×10^{-2}. In comparison, the visible meso-gamma scale features at reduced frequencies between $1\text{-}5\times10^{-3}$ were strongly enhanced in the high wind speed spectra. This is consistent with the previous findings [4, 7]. For the wind speeds encountered here, these are features with wavelengths between approximately one and five kilometers, and might be associated with boundary layer rolls.

Previous analyses of hurricane winds integral length scales, corresponding to the frequencies of maximum power in the power spectrum, ranged from less than hundred to several hundred meters for L_u, but were all observed over relatively flat terrain or with ocean fetches [2, 4, 6, 7]. Comparative length scale data to the one observed in Houston come from the Beijing tall tower measurements [33]. That study reported highly variable values with means of approximately 100 and 200 m for z=47 and 120 m agl. Similar to the data above, that data base contained only wind speeds up to 15 m s^{-1} at z=120 m, but the authors did not analyze for wind speed correlation. As integral length scales were found to increase with wind speed [7] it is possible that larger average length scales would have resulted had wind speeds been higher. Here, length scale data were calculated between noontime on DOY 256 and midnight under the limitations described above. The results are summarized in Table 1.

Length scale	**this study (mean ± sd)**		**Li et al. (means)**	**Yu et al. (mean ± sd)**	**Masters et al. (mean ± sd) 1**
	all data	**wake-free data**			
L_u	217±168	282±185	104, 195	122 ± 29	130 ± 45
L_v	177±137	214±153	44, 93		90 ± 43

[1]these values are averages over several landfalling hurricanes; the actual spread of the values for each hurricane was at times much larger than indicated by the sd of the means listed here

Table 1. Integral length scales in meters for winds <= 15 m s^{-1} (this study and [33], $z_0\approx1$ m), and >=15 m s^{-1} ([6] and [7], $z_0<0.1$ m)

The observed values are broadly comparable. Similar to all observations is the relatively large variability of the data, with several values exceeding 400-500 m length scales, and a generally log-normal distribution. A slight dependence on wind speeds and a good correlation between L_u and L_v was observed in this study's data and the data in [6]. The wind speed correlation suggests that values over 400 m should occur regularly for wind speeds exceeding 15 m s^{-1}, which is consistent with observations in [6]. The values over Beijing appear lower, but were likely influenced by a large amount of data at lower wind speeds, but this was not elaborated on [33].

4.3. Impacts

Sustained wind speeds and turbulence intensity account for the wind loads plotted in Figure 6. Typically, static loads exceed dynamic loads. However, dynamic wind loads can contribute significantly due to the increased turbulence created by wake effects of buildings in urban areas. Individual periods during which I_U was increased [17] locally were previously highlighted and discussed above. However, a single site measurement cannot assess all local effects. Where and when damages most likely occur has instead been assessed through standardized measurements and models in the past. For instance, the results from extensive works on wind-throw in forests in [30, 36, 37], [38], [39], and [31] may be applicable to urban areas because, as in Houston, urban tree distribution is often patchy. In addition, buildings represent additional roughness elements causing similar or stronger wakes than other trees and stands in parks. While wind loads in Houston (Fig. 6) were seemingly not high enough to uproot trees, another measure to gauge surface impacts is momentum transfer or, more generically, (surface) friction velocity, u_*.

Friction velocity was measured directly with the sonic anemometer, but also calculated by linearizing the log-law (equation 1) for 10-min section mean winds using the four measurement levels from 20-60 m agl. If the preset roughness length and displacement height values are correct, the linearization retrieves friction velocity from the slope of the curve and a near zero intercept under neutral and near neutral stabilities, including some allowance for error (e.g. ±1 for the intercept). The results are plotted in Figure 9. We found a generally excellent match between the sonic anemometer and the wind gradient values.

Biases were observed for three reasons: First, when strong instability is present, such as visible midday on DOY 255. Second, during periods when wind speeds at the lower levels are affected by the tower's wake, generally between approximately 320 and 10 degrees; u_* is first under-, then overestimated and a negative, then positive y-intercept in the linearization is observed. This can be explained by a temporary acceleration the wind undergoes approaching the sensor from close to the tower structure for approximately 0-10 deg. N. This increases the linearization slope and creates a negative intercept. As wind direction turns further N towards NNW, winds are slowed through the tower structure, which creates a lower slope and a positive intercept. In either case, large residuals and poor determination coefficients of the linearization resulted. In neither case were winds at the top level affected. Third, the strongest bias though was observed for NW wind directions during the latter part of the eyewall passage, marked by the vertical dashed line. Calculated u_* plummeted and large positive intercepts were found from the linearization. At the same time, lateral turbulence intensity, I_v, dropped strongly, see Fig. 7, while strong updrafts turned the saucer-shape radiation shields from their horizontal into a vertical position (pictures are available online). For these wind directions, the flow advects over the large industrial building in 100 m distance, which occupies a significant fraction of the footprint at these wind speeds and likely produces a significant wake and associated turbulence. If only the top level sensor were affected at this wind direction, the linearization determination coefficient would again have been low and residuals high. This was, however, not the case. Thus, surface characteristics for this wind direction are most likely not well described by the preset z_0 and d values for the encountered wind speeds.

http://dx.doi.org/10.5772/51344

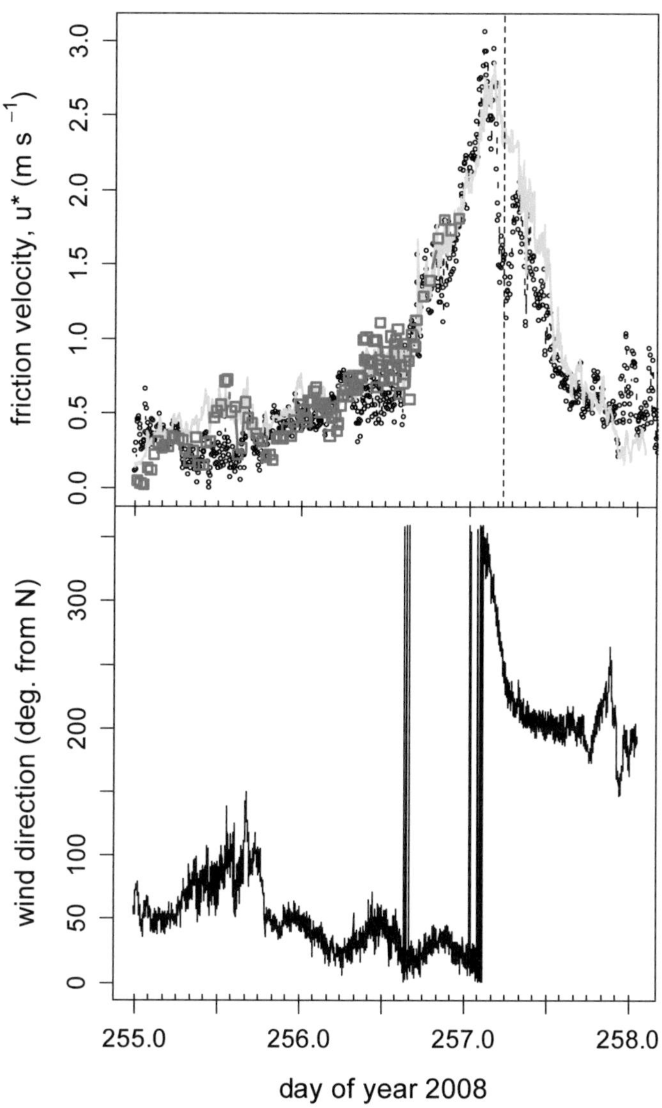

Figure 9. Friction velocity (upper panel) and wind direction (lower panel) time series during Hurricane Ike's passage over Houston. Small black circles depict all gradient-calculated values, open dark grey squares represent sonic-measured values, and the light grey line is a straightforward calculation of u_* using the log-law and only top level cup anemometer winds alongside preset z_0 and d. The vertical dashed line marks the period when winds advected from over the large building on the NW side of the tower.

In summary, Figure 9 shows extremely high, not previously measured friction velocities over an urban area. The corresponding large, turbulent downward momentum fluxes are the main destructive force in this environment. The results were observable after the storm. Figures 10 and 11 show the typical impacts in this neighborhood that led to the large scale power failures suffered all over the metro area.

Figure 10. Tree damage and debris along Hardy Road, ca. 2 km from the tower site.

Figure 11. View from the tower at 12 m agl towards NNW. The circles mark tree damage, the arrow on the lower left marks the local generator's exhaust pipe.

Two of the three tall trees immediately north of the tower, shown in Figure 11 growing alongside Hays Street between some of the larger buildings, and immediately downwind of the largest building in the area seen on the left, were also severely damaged, the one on the left losing nearly its complete crown.

It is also interesting that damage was largely confined to trees, or secondary damage from falling trees; significantly less (tile) roof damage was reported. This has previously been interpreted as being due to a mature tree canopy extending beyond roofs in many parts of Houston [40], with the trees taking the brunt of the storm's winds. However, in heterogeneous building areas such as the immediate surroundings of the tower site, small scale damage proportional to the encountered tropical storm force winds was caused. As shown in Figure 12, the local Yellow Cab headquarters suffered from a torn-off metal roof and a collapsed metal awning. In addition, not shown here, the company's antennas on top of the tower were bent and the tower's guy wires were loose, leading to significant sway of the tower. Similar damage kept cell-phone tower crews busy for months after the storm.

Another significant impact of hurricanes at landfall is the often torrential inland rainfall, leading to flash-floods and potential inundation of low lying areas. In this case, Houston was much more prepared, having learned lessons from many past floods, particularly tropical storm Allison in 2001. The next section will discuss Hurricane Ike's thermodynamic properties and the associated rainfall as compared to independent measurements.

Figure 12. Structural damage caused by Ike as viewed from the tower at approx. 18 m agl towards NE (top) and ESE (bottom)

5. Equivalent potential temperature and rainfall

Hurricanes have been called equivalents of heat engines [41]. A warm ocean in disequilibrium with the overlying (dry) air supplies large amounts of water vapor by evaporation at the ocean surface. When condensing in the atmosphere this energy is released as latent heat of vaporization during air flow towards a pressure minimum, and it is efficiently converted into kinetic energy, i.e. increasing wind speeds. Conservation of angular momentum forces these winds to spiral faster as they approach the low pressure center of the storm. Increased wind speeds also cause increased evaporation of surface ocean water, maximizing humidity towards the center of the storm, where often the most heavy precipitation occurs. The increase of entropy towards the storm's eye can be depicted by calculating equivalent potential temperature (Θ_e), the temperature an air parcel would assume when brought adiabatically to surface reference pressure (potential T) and using latent heat of vaporization from condensing all its available water vapor to heat it. The distribution of Θ_e for a mature, symmetric hurricane over a uniformly warm ocean surface is presumably also close to symmetric, increasing towards the eyewall, and observational studies tend to confirm this structure, e.g. [42]. When hurricanes make landfall, their source of (latent heat) energy at the surface is cut off and drier air is eventually imported into the storm, leading to its inevitable dissipation over land. At a single measurement location, the 2-D thermodynamic structure of the hurricane can be observed via Θ_e when it moves past the sensor. Here, Ike's track allowed for this observation from the time of its approach to the coastline to its passage into northeast Texas. In Figure 13, Θ_e is plotted alongside measured precipitation and wetness. The wetness sensor typically allows for the detection of light rain when there was not enough rainfall to be detected by the tipping rain bucket instrument.

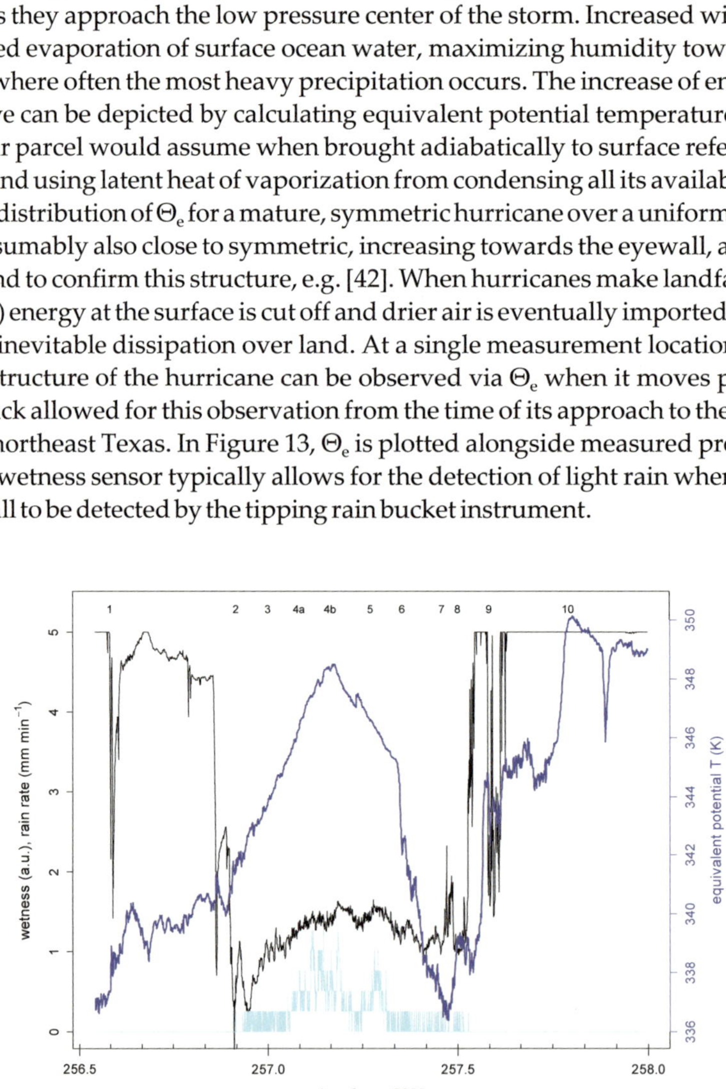

Figure 13. Time series of Θ_e (thick blue line), precipitation (light blue line), and wetness (thin black line) during Hurricane Ike's passage over Houston (wetness is plotted as the logarithm of the raw sensor signal, a resistance measurement). Similar to previous plots, radar-based evaluation of rain band passage timing is plotted along the top with bands 4a and 4b representing the eyewall.

It is clear from Fig. 13 that Θ_e near monotonously increased during the storm's approach and into its eyewall. Rapid increases were observed in the outer rain bands, including bands 8 to 10 in the afternoon after landfall, when the storm's center had migrated into northeast Texas and local precipitation had all but ceased. Increases in humidity as indicated by the wetness sensor were often clearly related to increases in Θ_e. Significant precipitation began to fall at the site shortly before 22:00 CST, and increased steadily into the eyewall. The highest rain rates were observed in the NW side of the eyewall, coincident with the highest wind speeds under still rising Θ_e levels (03:00 CST), and on the SW side of the eyewall just after the lowest pressure had been observed (04:00 CST). Then Θ_e rapidly dropped, and despite a short-term increase in an outer rain band with heavy precipitation (#5) fell to pre-storm levels before midday on DOY 257. The following afternoon rise, though associated with outer rain bands of the storm, was not associated with strong winds or significant precipitation any more.

This development, including its correlation with the winds as shown in Fig. 5, can be interpreted through the air circulation into the storm during and after landfall. For a straightforward elucidation, we carried out both a back trajectory analysis using HYSPLIT [43, 44] with 40 km resolution EDAS data, and an analysis of locally measured carbon monoxide concentrations as a tracer of continental air masses. In the first hours after landfall up to the approach of the eye closest to central Houston, the observed air masses were tropical marine in origin. As the storm approached the site it "imported" its Θ_e onto land. Beginning at approximately 04:00 CST, when the eye lay east of the tower site, the storm began to entrain drier continental air from north and northwest Texas into its western eyewall, which can explain the rapid decline in Θ_e. Figure 14 shows two back trajectory ensembles, one for just before the Θ_e decline, and one for six hours after decline had begun. This entrainment explains why Ike maintained much stronger wind speeds particularly on its eastern side throughout the morning [28] as this side continued to be fueled by Gulf of Mexico air masses. With the storm moving further north by midday and into the afternoon, wind direction had changed back to southerly flows with all air mass origins returning to the Gulf of Mexico (data not shown). This analysis is corroborated by the carbon monoxide data shown in Figure 15.

Typical continental boundary layer CO levels for September of 100-130 ppb were measured on DOY 256, with higher values during the morning and afternoon rush hour periods. As the storm approached and a tropical marine air mass was established, CO dropped to northern hemisphere marine background levels near 75 ppb, among the lowest levels measured in Houston. However, shortly after Θ_e levels began to drop and wind speeds plummeted, CO levels rose again to levels consistent with entraining continental air. Thus, the west side of the storm was rapidly deprived of its latent heat energy source. Nevertheless, short-term, small fluctuations in CO abundance, anti-correlated with Θ_e, could still be observed during the early afternoon of DOY 257. Later on, the reason for continued CO increases even after air mass origins had fully returned to tropical marine in the afternoon of DOY 257 can be found in a low nighttime boundary layer and large local CO emissions from electricity generators in the Houston metro area [45], including Yellow Cab's own onsite diesel generator (see Fig. 12).

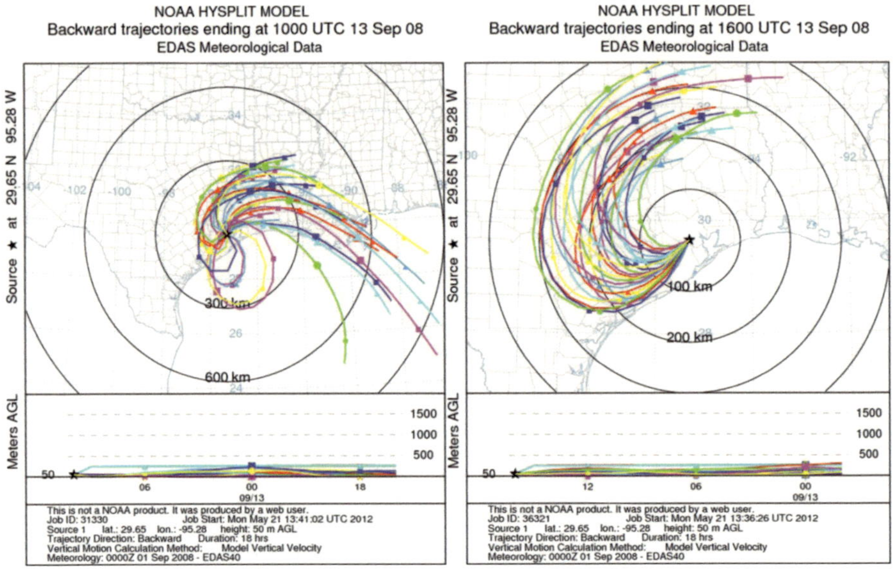

Figure 14. Back trajectory ensembles for 04:00 CST (left) and 10:00 CST (right) on DOY 257.

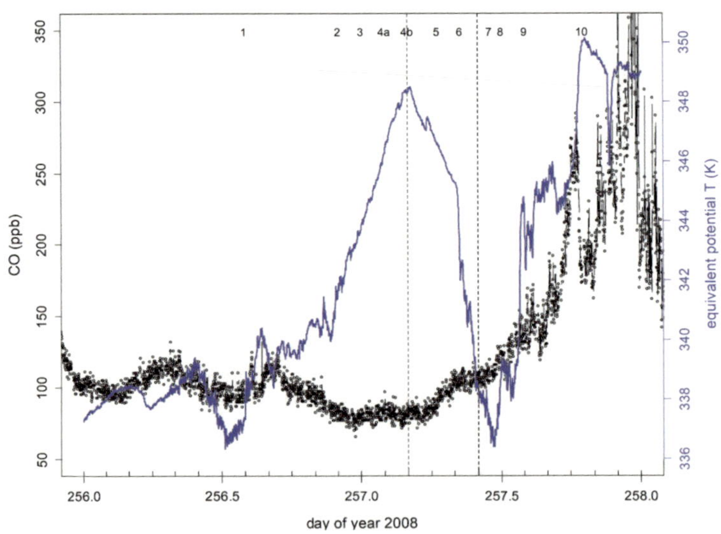

Figure 15. Time series of carbon monoxide (CO) levels as compared to Θ_e prior and during Ike's passage over Houston. As in previous Figures, rain band passage timing is indicated along the top. The two vertical dashed lines mark the start times for the back trajectories plotted in Fig. 14.

In addition to Θ_e rainfall dynamics were analyzed. A time series analyses of the storm data showed principle periodicities of 20 and 50 minutes, corresponding to rain band recurrence. For the flooding potential, the total amount and intensity of rainfall is of particular interest.

As Houston is flood-prone, several rain and flood gauges are maintained throughout the water sheds in the metro area. Thus, the locally measured rainfall was compared to the collection of data provided online [28] in Figure 16.

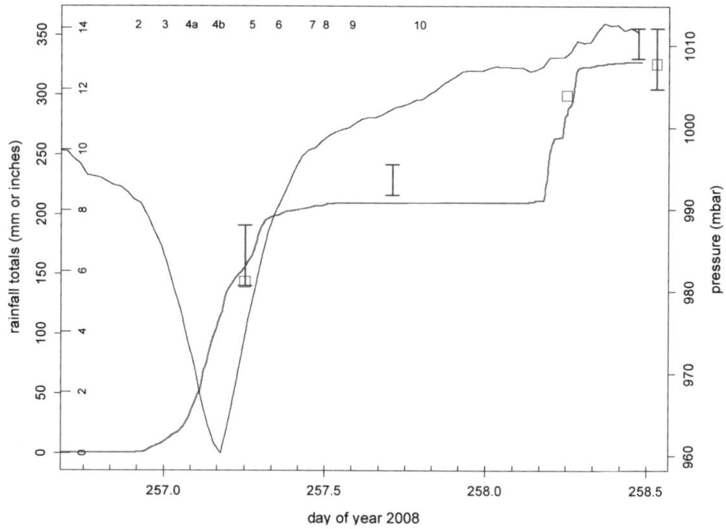

Figure 16. Rainfall development (blue line) during both Hurricane Ike's and the subsequent day's morning frontal passage over Houston. Vertical dark blue bars mark radar-derived precipitation totals, open squares mark totals observed at the nearest USGS gain gauge in the White-Oak-Bayou water shed. Pressure and rain band passage timing are plotted for orientation.

In general, a good agreement was found between the extrapolated radar data and the closest rain gauge to the tower. There appears to be a slight underestimation of the local measurements as compared to the references, which could be explained by the location of the rain gauge on the tower: As it is installed on the east side of the tower, rainfall approaching from the west and southwest during the second half of the eyewall passage may not have all reached the bucket. However, the highest rainfall rates were recorded during the NW eyewall passage consistent with the radar data [28], reaching 11 mm $(10\ \text{min})^{-1}$. Nearly twice the intensity, 19 mm $(10\ \text{min})^{-1}$, however, was observed during the post-storm frontal passage on DOY 258 fueled by storm-intensified humidity entrainment from the Gulf of Mexico [28], Fig. 16, which added almost as much rainfall to the total as caused by Ike during the previous day. Ultimately, a total of 327 mm (12.9 inches) of rain fell onsite within 36 hours, with up to 15 inches as estimated from the radar data for other areas of Houston [28], more than three times the long term average September rainfall for this area [46].

6. Conclusions

Hurricane Ike's path after landfall during the early morning hours on 13 September 2008 crossed through Galveston Bay, went north along the eastern border of Harris County, then slightly northwest again towards north of Houston, until finally veering NNE towards northeast Texas. Most of Harris County, including the city of Houston and the economically important Ship Channel, went through the western, often less severe eyewall of the hurricane. Nevertheless, the Houston metro area suffered from widespread power outages caused by the hurricane's impacts hours before the highest rain rates and wind speeds passed. Onsite measured rainfall compared well with independent data. While local flooding was caused by this three-times monthly average amount in 36 hours, the Houston drainage network including its improvements since tropical storm Allison in 2001 seemingly handled the amount of water adequately [47]. The same was not the case with respect to preparations for potential wind damage. This chapter described Ike's wind field from approach to retreat as measured from a tower platform north of downtown Houston. Measured winds were likely representative of the larger metro area, showing tropical storm force winds at 20 m agl, strongly slowed by the rough urban surface as compared to standardized H*Wind products. However, although measured wind speeds were significantly lower than commonly cited, strong downward momentum fluxes in combination with high turbulence intensities, likely locally exaggerated by building wakes, caused "moderate" tree and building damage in the area equivalent to what would be expected from a category two hurricane. As this "moderate" damage dominantly occurred to a mature, poorly trimmed tree canopy in neighborhoods where power is distributed aboveground on poles, much higher than expected power line loss resulted.

Similar to previous measurements, the wind analysis results also indicate larger horizontal structures in the wind field not present at lower wind speeds. Significant interaction of these *roll* features with the underlying surface might contribute to the damage potential of the storm, but this cannot be evaluated with confidence from a single observing location. The extensive meteorological observations along the Gulf Coast during Hurricane Ike may, however, provide the opportunity for such an analysis once all data become analyzed in more detail.

Additional observations of equivalent potential temperature and air quality development during the approach and passage of Hurricane Ike revealed that Houston may have been spared even larger damage from high winds because of entrainment of drier continental air into the western eyewall. While higher wind speeds were maintained south and east of the eye, Houston in the west of the eye experienced overall lower winds and shorter durations of tropical storm force winds than observed on the east side of the eye [28].

Acknowledgements

I am indebted to the employees of Houston Yellow Cab at 1406 Hays Street, particularly William Hernandez, who were onsite during the storm to keep the power from their generator on and protect our equipment against water that had penetrated the flat roof structure shown in Fig. 12. The Greater Houston Transportation Co. has provided free access to its tower and base building for our study since 2007, thereby facilitating and supporting these unique urban measurements. We continue to be grateful for this private enterprise support of our academic goals. I also thank my graduate student Marty Hale for assembling Figure 4b. Our research

http://dx.doi.org/10.5772/51344

was supported at the time by the Texas Air Research Center (TARC), Lamar University, Beaumont, TX, and a start-up grant to the author by Texas A&M University in College Station.

Author details

Gunnar W. Schade

Address all correspondence to: gws@geos.tamu.edu

Department of Atmospheric Sciences, Texas A&M University, College Station, TX, USA

References

[1] Pielke, R.A., J. Gratz, C.W. Landsea, D. Collins, M.A. Saunders, and R. Musulin, Normalized hurricane damage in the United States: 1900-2005. Natural Hazards Review, 2008. 9(1): p. 29-42.

[2] Lorsolo, S., J.L. Schroeder, P. Dodge, and F. Marks, An Observational Study of Hurricane Boundary Layer Small-Scale Coherent Structures. Monthly Weather Review, 2008. 136: p. 2871.

[3] Powell, M.D., S.H. Houston, and T.A. Reinhold, Hurricane Andrew's Landfall in South Florida. Part I: Standardizing Measurements for Documentation of Surface Wind Fields. Weather and Forecasting, 1996. 11: p. 304-328.

[4] Schroeder, J.L. and D.A. Smith, Hurricane Bonnie wind flow characteristics as determined from WEMITE. Journal of Wind Engineering and Industrial Aerodynamics, 2003. 91: p. 767-789.

[5] Skwira, G.D., J.L. Schroeder, and R.E. Peterson, Surface Observations of Landfalling Hurricane Rainbands. Monthly Weather Review, 2005. 133: p. 454-465.

[6] Masters, F.J., H.W. Tieleman, and J.A. Balderrama, Surface wind measurements in three Gulf Coast hurricanes of 2005. Journal of Wind Engineering and Industrial Aerodynamics, 2010. 98(10-11): p. 533-547.

[7] Yu, B., A.G. Chowdhury, and F.J. Masters, Hurricane Wind Power Spectra, Cospectra, and Integral Length Scales. Boundary-Layer Meteorology, 2008. 129(3): p. 411-430.

[8] Doran, J.C., J.D. Fast, and J. Horel, The VTMX 2000 campaign. Bulletin of the American Meteorological Society, 2002. 83(4): p. 537-551.

[9] Chen, F., R. Bornstein, S. Grimmond, J. Li, X. Liang, A. Martilli, S. Miao, J. Voogt, and Y. Wang, Research priorities in observing and modeling urban weather and climate. Bulletin of the American Meteorological Society, 2012(in press).

[10] Changnon, S.A., INADVERTENT WEATHER-MODIFICATION IN URBAN AREAS - LESSONS FOR GLOBAL CLIMATE CHANGE. Bulletin of the American Meteorological Society, 1992. 73(5): p. 619-627.

[11] Berkowitz, C.M., S. Springston, J.C. Doran, J.D. Fast, and Ams, Vertical mixing and chemistry over an arid urban site: First results from aircraft observations made during the Phoenix sunrise campaign. Fourth Conference on Atmospheric Chemistry: Urban, Regional and Global Scale Impacts of Air Pollutants2002. 165-168.

[12] Allwine, K.J., J.H. Shinn, G.E. Streit, K.L. Clawson, and M. Brown, Overview of urban 2000 - A multiscale field study of dispersion through an urban environment. Bulletin of the American Meteorological Society, 2002. 83(4): p. 521-536.

[13] Oke, T., K. Klysik, and C. Bernhofer, Editorial: Progress in urban climate. Theoretical and Applied Climatology, 2006. 84(1-3): p. 1-2.

[14] Oke, T.R., Towards better scientific communication in urban climate. Theoretical and Applied Climatology, 2006. 84(1-3): p. 179-190.

[15] Oke, T.R., Initial guidance to obtain representative meteorological pbservations at urban sites, 2006, World Meteorological Organisation: Geneva.

[16] Roth, M., Review of atmospheric turbulence over cities. Quarterly Journal Of The Royal Meteorological Society, 2000. 126(564): p. 941-990.

[17] Schade, G.W., Relating urban turbulence and trace gas flux measurements from a tall tower to surface characteristics and anthropogenic activities, 2009, Texas A&M University: College Station. p. 42.

[18] Park, C., G.W. Schade, and I. Boedeker, Flux measurements of volatile organic compounds by the relaxed eddy accumulation method combined with a GC-FID system in urban Houston, Texas. Atmospheric Environment, 2010. 44(21-22): p. 2605-2614.

[19] CenterPoint Energy Inc. Hurricane Ike, Like many things, bigger in Texas. 2009 [cited 2012, 5 July]; Available from: http://www.centerpointenergy.com/newsroom/stormcenter/ike/.

[20] Berg, R., Tropical Cyclone Report, Hurricane Ike, 1 - 14 September 2008, 2009/2010, National Hurricane Center. p. 55.

[21] Hurricane Ike Impact Report, FEMA, Editor 2008, ESF#14. p. 64.

[22] G.W., S. and B. Rappenglueck, Unique Meteorological Data During Hurricane Ike's Passage Over Houston. EOS, 2009. 90(25): p. 215-216.

[23] Fowler, T., Next hurricane may not knock out Houston power so long, Lights could be restored faster, Electric utilities apply lessons of Hurricane Ike, in Houston Chronicle 2009, Hearst Communications Inc.: Houston, Texas.

[24] Fowler, T., Lesson of Ike: Money needed for power grid vs. blackouts, Investment in grid could bring quicker power return, Task force wants new technology it says can

http://dx.doi.org/10.5772/51344

cut wait time after storm outages in Houston Chronicle 2009, Hearst Communications Inc.: Houston, Texas.

[25] Park, C., G.W. Schade, and I. Boedeker, Characteristics of the flux of isoprene and its oxidation products in an urban area. Journal Of Geophysical Research-Atmospheres, 2011. 116.

[26] Grimmond, C.S.B., T.S. King, M. Roth, and T.R. Oke, Aerodynamic roughness of urban areas derived from wind observations. Boundary-Layer Meteorology, 1998. 89(1): p. 1-24.

[27] Chambers, M., The Hurricane Severity Index – A New Method of Classifying the Destructive Potential of Tropical Cyclones, in Severe Storm Prediction and Global Climate Impact on the Gulf Coast2008, SSPEED: Rice University, Houston, TX.

[28] National Weather Service (NWS) Houston/Galveston. Hurricane Ike (2008). 2012 [cited 2012 10 June]; Available from: http://www.srh.noaa.gov/hgx/?n=projects_ike08.

[29] HGAC, 2011 Regional Storm Debris Management Assessment, 2011, Houston-Galveston Area Council (H-GAC). p. 155.

[30] Flesch, T.K. and J.D. Wilson, Wind and remnant tree sway in forest cutblocks.: I. Measured winds in experimental cutblocks. Agricultural and Forest Meteorology, 1999. 93(4): p. 229.

[31] Panferov, O. and A. Sogachev, Influence of gap size on wind damage variables in a forest. Agricultural and Forest Meteorology, 2008. 148(11): p. 1869.

[32] Powell, M.D., New findings on hurricane intensity,wind field extent, and surface drag coefficient behavior, in Tenth international workshop on wave hindcasting and forecasting and coastal hazard symposium2007: North Shore, Oahu, Hawaii. p. 14.

[33] Li, Q.S., L.H. Zhi, and F. Hu, Boundary layer wind structure from observations on a 325 m tower. Journal of Wind Engineering and Industrial Aerodynamics, 2010. 98(12): p. 818-832.

[34] Varshney, K. and K. Poddar, Experiments on integral length scale control in atmospheric boundary layer wind tunnel. Theoretical and Applied Climatology, 2011. 106(1-2): p. 127-137.

[35] R Development Core Team. R: A language and environment for statistical computing. 2012; Available from: http://www.R-project.org/.

[36] Flesch, T.K. and J.D. Wilson, Wind and remnant tree sway in forest cutblocks. II. Relating measured tree sway to wind statistics. Agricultural and Forest Meteorology, 1999. 93(4): p. 243.

[37] Wilson, J.D. and T.K. Flesch, Wind and remnant tree sway in forest cutblocks. III. a windflow model to diagnose spatial variation. Agricultural and Forest Meteorology, 1999. 93(4): p. 259.

[38] Zeng, H., H. Peltola, A. Talkkari, A. Venäläinen, H. Strandman, S. Kellomäki, and K. Wang, Influence of clear-cutting on the risk of wind damage at forest edges. Forest Ecology and Management, 2004. 203(1-3): p. 77.

[39] Dupont, S. and Y. Brunet, Simulation of turbulent flow in an urban forested park damaged by a windstorm. Boundary-Layer Meteorology, 2006. 120(1): p. 133-161.

[40] WeatherPredict Consulting Inc., Hurricane Ike Meteorological Assessment and Damage Survey, Sptember 2008, RenaissanceRe, Editor 2008.

[41] Emanuel, K., Hurricanes: Tempests in a greenhouse. Physics Today, 2006. 59(8): p. 74-75.

[42] Bell, M.M. and M.T. Montgomery, Observed structure, evolution, and potential intensity of category 5 Hurricane Isabel (2003) from 12 to 14 September. Monthly Weather Review, 2008. 136(6): p. 2023-2046.

[43] Draxler, R.R. and G.D. Rolph. HYSPLIT (HYbrid Single-Particle Lagrangian Integrated Trajectory). 2012; Available from: http://ready.arl.noaa.gov/HYSPLIT.php.

[44] Rolph, G.D. Real-time Environmental Applications and Display sYstem (READY). 2012; Available from: http://ready.arl.noaa.gov.

[45] George, C., Ike had hidden toll in gaming kids sickened by generator, in Houston Chronicle 2009, Hearst Communications Inc.: Houston.

[46] Office of the State Climatologist. 2012 [cited 2012, 5 July]; Available from: http://atmo.tamu.edu/osc/tx.

[47] Berger, E., Progress and lessons 10 years after Tropical Storm Allison, in Houston Chronicle 2011, Hearst Communications Inc. : Houston.

Section 3

Preparedness and Impacts

Chapter 6

Application of Simulation Modeling for Hurricane Contraflow Evacuation Planning

Gary P. Moynihan and Daniel J. Fonseca

Additional information is available at the end of the chapter

http://dx.doi.org/10.5772/50976

1. Introduction

The Gulf Coast and Atlantic coastal states of the U.S. are often subjected to severe tropical storms and hurricanes. Hurricane season nominally extends from June 1 to November 30 of each year. From 1851 to 2006, there have been 279 landfalls on the mainland U.S. coastline, including 96 major hurricanes of Category 3 and above. Among these, the thirty most costly strikes resulted in an estimated total loss of approximately $346 billion in 2006 dollars, and took more than 19000 lives [1]. The usual response to these severe weather events is to evacuate inland from the coast. Normal traffic flows may turn into congestion, frustration and gridlock. This reduces the number of vehicles that can leave the coastal area if an evacuation order is issued. The potential risk for loss of life increases if the hurricane strikes stalled traffic, as people's efforts to evacuate might place them at greater risk than they would have faced if they had stayed put. In response to Hurricane Floyd (see Figure 1), extensive traffic delays occurred along inland evacuation routes throughout the state of South Carolina. Subsequently, The U.S. Federal Emergency Management Administration (FEMA) conducted regional meetings to identify approaches for better traffic planning, management, and coordination. These planning efforts have continued at the federal, regional, state, and local levels.

With modern weather forecasting techniques, the path and associated strength of an advancing hurricane can be predicted with some confidence. Progressively more attention has been paid to improving the planning and operations of hurricane evacuation to reduce unnecessary losses in the projected area of landfall or near landfall. Evacuation planning for a large area frequently involves multiple considerations, e.g. shelter location, evacuation routes, flow assignment, allocation of emergency response and law enforcement resources. Operational strategies may include real time traffic monitoring, advanced traveler alerts, signal timing adjustment for local arterials, and rerouting both local and interstate roads. In the United States, several Southeastern states have adopted the concept of "contraflow", or "reverse-laning", in hurricane evacuation. "Reverse-laning" is the process of reversing one direction of traffic on specific routes to facilitate overall evacuation flow. This procedure is generally applicable to interstate roadways, referred to as "denied access routes", since traffic control can be applied to interchanges and terminal areas. (See Figure 2) The idea is to reverse one direction of the roadway in order to accommodate the often substantially increased travel demand moving away from the impact area. Actual implementation of reverse-laning varies from state to state. For example, in the states of Texas and Florida, each county or regional area has its own evacuation plan, many of them involving contraflow. Contraflow operations are only executed in Texas if a mandatory evacuation order is issued by the respective mayor or county judge [2]. Reverse-laning plans for the major metropolitan areas are detailed by the Texas Department of Transportation [3]. In the event of voluntary evacuations, there is no actual lane reversal and the shoulders of the road are used as travel lanes. In Florida, the State's Department of Transportation coordinates the individual counties evacuation plans, such that the following roadways utilize contraflow for evacuation: I-75, I-10, I-4, the Florida Turnpike, and State Road 528 [4]. Louisiana and Mississippi share a unique plan for shared hurricane evacuation. Because of its small coastal population, Mississippi does not utilize contraflow within its own borders, since its roadways can handle evacuation traffic without modification. The city of New Orleans in Louisiana is a major population center, whose evacuation routes may go through Mississippi. The two states thus coordinate their contraflow operations to avoid confusion and disruption [5].

Figure 1. Hurricane Floyd approaching the South Carolina coast (September 1999)

Figure 2. Traffic control using temporary barriers during contraflow operations

During the past 60 years, 25 hurricanes have made landfall along the Gulf Coast. Of those, five (Hurricanes Frederick, Eloise, Opal, Ivan, and Dennis) have had the eye of the hurricane make landfall in Alabama [6]. Many other hurricanes, that may have not made actual landfall in Alabama, have caused significant damage (e.g. Hurricane Katrina in 2005). As an increasing number of Alabama's population lives in the eight counties closest to the Gulf of Mexico (See Table 1), it is becoming progressively more vulnerable to these extreme weather events [7]. The Alabama Department of Transportation (ALDOT) has developed a well-planned evacuation procedure for this coastal area of the state. In this plan, an approximately 140-mile section of Interstate roadway 65 (I-65) between exit 31 and exit 167 is identified as the contraflow segment. As noted in Figure 3, I-65 would be reverse-laned such that all traffic would flow north, from south of Alabama Route 225, near the large population center of Mobile, to Exit 167/168, just south of the greater Montgomery metropolitan area [8]. This concept was further refined after Hurricane Katrina, to provide emergency vehicles with an alternate route south via U.S. 31.

ALDOT's reverse-laning plan identifies four operating levels [8]. Level 1 begins with the start of each hurricane season. Level 2 is initiated when the U.S. National Weather Service (NWS) issues a hurricane watch for the Gulf Coast of Alabama, Mississippi, and the "panhandle" region of northwest Florida. Level 3 is indicated when the NWS watch is upgraded to a hurricane warning. During these first three operating levels, the required equipment are gradually staged and personnel prepare for contraflow operations. The actual reverse-laning occurs during Level 4, when the State's Transportation Director (in consultation with the Alabama Emergency Management Services) issues the order for contraflow, and extends until he orders termination of the operations [8].

County	Population	Persons/Vehicle
Mobile	400526	.96
Washington	17906	.73
Baldwin	156701	.73
Escambia	38336	.96
Conecuh	13453	.86
Monroe	23725	.96
Butler	20764	1.21
Lowndes	13210	1.21

Table 1. Alabama counties included in this study

Figure 3. Map of contraflow segment in Alabama

2. Problem statement

The current practice in Alabama is a staged process. Equipment and personnel are deployed first, and the actual call for reversing the southbound lanes depends on the measured traffic condition and other relevant factors [8]. Although these contraflow operations increase the roadway capacity for evacuation, this reverse-laning is, by its necessity, a unique measure requiring extraordinary efforts. Practical implementation issues include traffic control, access management, use of roadside facilities, safety, labor requirements, and cost [9]. Therefore, care must be taken in the planning and real-time operations of contraflow evacuation. However, such decisions are often made in an ad hoc manner during actual implementation. Review of the literature indicates that a number of simulation frameworks [10-12] as well as several optimization models for evacuation flow assignment [13-17] have been developed to assist decision-makers during in emergency evacuations. However, the literature further indicates that few studies have directly addressed specific issues related to contraflow planning and operation. Kim et al. [18] and Lv et al. [19] try to determine which lanes in a transportation network should be reversed from a system perspective. Theodoulou and Wolshon [20] and William et al. [21] focus on detailed configurations of the starting point of the specified contraflow segment. Meng and Khoo [22] consider the onset and duration of contraflow in an integrated problem.

Selection of a suitable evacuation model is requisite to support the needs of the Alabama Department of Transportation Maintenance Bureau regarding their responsibilities for contraflow planning and evaluating possible responses to a hurricane event. The I-65 evacuation route has been subject to considerable analysis [23, 24]. Yet these analyses have focused only on capacity and congestion issues relating to I-65, itself. It has been suggested that selective control of specific on-ramps may improve the effectiveness of the overall evacuation routes. For example, prioritization could be based on level of danger (giving people living closest to the coast priority access to I-65, with other communities directed to other state roads). These planning alternatives could be evaluated through an improved evacuation planning model.

3. Research course

Review of the literature has noted considerable work on the development of decision rules and computer-based support systems to aid in decision-making [25]. A variety of mathematical models have been developed which focus on evacuation route planning. These network models of evacuation problems are extensions of the classical operations research assignment problem. For these problems, the basic form of the network is that of the more general minimal cost transshipment (or flow) network. In the network, the arcs represent the flow of people, the source nodes represent initial source inventories (points of entrance into the evacuation network), and the sink nodes represent the final inventories (in this case, destinations). Optimization models (e.g. linear programming, goal programming or dynamic programming) are another category of mathematical models. The model is formulated to either maximize (or minimize) the objective function (depending upon the purpose of the model) within the context of available resources and constraints [26]. Coastal hurricane evacuation can be seen as a network optimization problem aimed at selecting the "best" routes from a set of candidate roads within an existing roadway network. This selection involves determining where the potential evacuation routes' origin and destination points are located, their maximum traffic volumes, and the type of evacuation schemes resulting in a maximum vehicle exit rate with minimal travel times [27-30].

A significant issue in managing a disaster evacuation operation is the pattern of flow of the roadways, i.e., equilibrium or non-equilibrium flow. The equilibrium network is satisfied when the distribution of flow in the network follows Wardrop's stated principles [31-33]. These principles note that the total flow of evacuating vehicles eventually reaches an equilibrium state in which every car has the same travel time. Conversely, a flow pattern in a non-equilibrium flow model cannot satisfy these flow constraints since each vehicle uses a distinct route depending upon the overall evacuation strategy utilized [34]. The network evacuation problem can be further categorized as either discrete or continuous network repetitions. Discrete network analysis emphasizes the search of evacuation scenarios in terms of capacity enhancements [34] The objective of the analysis is to select those roads to be included in the evacuation network, incorporating the effects that such a decision may have on the volume of traffic leaving the area under distress.

4. Application area

Continuous network modeling focuses on maximizing the capacity expansion of existing, predetermined evacuation networks. Monte Carlo simulation via discrete simulation was originally considered for projecting the uncertainties in traffic flow during the study. When analyzing highly congested highways, or super-saturated conditions, consideration of constructing a new alternative road would seem reasonable. Unfortunately, this approach could lead to Braess's paradox, in which case, the vehicles on the existing highway and the new road would travel much slower than before. The cited paradox was discovered in 1968 by Dietrich Braess, and was originally developed regarding the congestion of signals in transmission networks [35]. Braess determined that increased capacity in congested electronic networks slows down communication. During the past decades, several authors have focused their efforts in understanding the implications of Braess's paradox [32, 35-37]. They have developed heuristics and mathematical models to predict and explain why this counterintuitive situation occurs. However, this knowledge has not been used by highway traffic planners. There is evidence of Braess's paradox in newly constructed roads around the world, as in the case cited of a road built in Stuttgart, Germany, which deteriorated traffic conditions to a point where it had to be closed down [37].

The major obstacle for the application of the mentioned heuristics and mathematical models to improve traffic conditions in congested highways, as is the case during a massive evacuation event, is the lack of knowledge on the premises of Braess's paradox. Subsequent work by Fonseca et al. [38] demonstrated that this could be extended to traffic analysis in a small city. Investigation was conducted regarding the further application of this approach to better project traffic congestion due to hurricane evacuation from the Gulf Coast. The main purpose of this study is to create a prototype model, following Braess's paradox premises, for improved hurricane evacuation planning.

http://dx.doi.org/10.5772/50976

5. Method used

Probabilistic models analyze the natural variation of conditions, as opposed to determining a mathematical optimum. For example, Fu and Wilmot [39] applied a logit model to estimate the conditional probability of households evacuating during a given time period prior to hurricane landfall. Many conventional algorithmic models may not sufficiently apply to specific domain problem areas, e.g. traffic planning. The utilization of computer-based simulation is a frequent means of probabilistic modeling, as well as a well-accepted approach of modeling complex systems and activities. A simulation model is primarily mathematical in nature. Rather than directly describing the overall behavior of the system under investigation, the simulation model attempts to "replicate" this behavior by studying the interactions among its components. The system is divided into elements whose behavior can be predicted in terms of probability distributions, for each of the various possible states of the system and its inputs [26]. Model output is normally presented in terms of selected metrics that reflect the performance of a system. Simulation has many advantages. It can provide a complete view of the total operations flow. Perhaps the most important advantage of a simulation is that it provides the opportunity for what-if analysis; i.e. it can project the impact of factors under a variety of conditions. The various decision alternatives then may thus be evaluated economically without disrupting existing operations, or incurring unnecessary costs.

During a hurricane evacuation, where the massive flow of vehicles takes place within a relatively short span of time, the traffic network moves from a situation of over-congestion to over-saturation. Under over-saturation conditions, traffic flow optimization is not feasible due to the overwhelming network inflow rate as compared to the exiting rate; thus, the utilization of computer-based simulation is a more appropriate means of modeling the complexity of the flow pattern. The evacuation network involved is this study corresponds to a discrete network presenting a flow pattern in equilibrium. Different levels of traffic representations are used by different classes of simulation models. In microscopic simulation models, the interactions of individual vehicles "are captured by using algorithms that represent vehicle acceleration and deceleration, passing maneuvers, and lane changing behavior" [40]. Tanaka [41] provides such a microscopic simulation model of hurricane evacuation on a single lane highway. The amount of detail, that is required at the microscopic level, becomes overwhelming when trying to model a large-scale evacuation over a large area. Macroscopic or mesoscopic models are favored under such circumstances. Macroscopic models are used to simulate traffic flow based on speed and traffic density relationships, and do not model the interactions between individual vehicles. Mesoscopic simulation models address individual vehicles in the transportation system, but capture their relationships using aggregate relationships. To enhance the modeling capability for real evacuation events and help the decision-maker in contraflow planning, this research will investigate the macroscopic approach, and develop a proof of concept simulation tool.

6. Status

The network consists of a single major US Interstate highway (I-65) with a set of 20 available on-ramp exits, and eight associated counties. The closing or opening of these ramps to traffic bound north represents the main decision variable of the analysis. Accepted development methodology identifies five primary phases: 1) data acquisition, 2) system design, 3) system construction, 4) verification and validation, and 5) experimentation and analysis. During the data acquisition phase, the key concepts and relationships were identified. Although the focus of this effort was on the development and evaluation of an evacuation planning simulation model for I-65, investigation also identified a large number of literature articles devoted to individual and collected hurricane evacuation case studies. These included the "Alabama Hurricane Evacuation Study" [42], and the National Oceanic and Atmospheric Administration's Hurricane Planning and Evacuation Assessment Reports [43]. These references emphasize best practices and lessons learned from a number of hurricane evacuations, as opposed to identifying specific algorithmic models for planning purposes. They did prove valuable as sources of data. Both the "Alabama Hurricane Evacuation Study – Summary Report", and the related "Alabama Hurricane Evacuation Study – Transportation Study [42, 44] were particularly useful in detailing issues regarding evacuation behavior. In addition to the literature search, the ALDOT traffic database was interrogated to obtain traffic data regarding interstate and state highways. Table 2 shows the estimates of the traffic volumes emerging from each on the 20 selected ramps during an eventual hurricane evacuation.

Exit Number	Exit Location	Total Vehicels
22	Washington Mobile	441743
31	Baldwin	53664
34	Baldwin	53664
37	Baldwin	53664
45	Baldwin	53664
54	Escambia	9983
57	Escambia	9983
69	Escambia	9983
77	Escambia	9983
83	Conecuh Monroe	10112
93	Conecuh Monroe	10112
96	Conecuh Monroe	10112
101	Conecuh Monroe	10112
107	Butler	4307
114	Butler	4307

Exit Number	Exit Location	Total Vehicels
128	Butler	4307
130	Butler	4307
142	Lowndes	3654
151	Lowndes	3654
158	Lowndes	3654

Table 2. I-65 exit locations and respective traffic levels

This led to the development of an overall proof of concept simulation model, initially focusing on management of entry ramps, applied to selected areas of the I-65 extended network. The evacuation model was constructed utilizing the discrete simulation software, *Arena*, which was interfaced with *Excel* macros for improved data input processing. Arena is a software product of Rockwell Automation, and combines both high-level modeling and general-purpose procedural programming. The software incorporates interchangeable templates of graphical simulation objects and statistical data analysis modules [45]. The discrete simulation model assesses the effect of closing selective ramps on the overall traffic evacuation rate, i.e., the number of vehicles evacuated from the area in question per hour. Consistent with Braess' Paradox, it was hypothesized that having all ramps open to vehicles exiting the region might actual be detrimental for the overall evacuation effort since in-flow congestion may be generated by entry ramps located within a few miles from each other; and due to the difficulty encountered by emergency and law-enforcement vehicles bound south when all ramps are exclusively for north-bound traffic. The models of each devised scenario were run at least 30 times to ensure the correctness of the statistical analyses performed.

The resulting simulation, for hurricane evacuation of inhabitants in the vicinity of the City of Mobile, Alabama, consists of a system of one top-level and two supporting models (see Figure 4). The top-level model is based on the entry of vehicles from the 20 on-ramps to I-65. The two supporting models assist the primary model with related traffic events such as car breakdowns and accidents, traffic control measures, inter-arrival signaling, and unforeseen emergency incidents. In the top-level model, entering vehicles are created through a controlled wait-and-signal mechanism [45]. (See Figure 5.) Attributes such as time of arrival, final destination exit, and accident incident proneness are established. These attributes are assigned based on cumulative probability distributions generated by empirical data collected during the data acquisition phase of this research project. Whenever a vehicle enters the highway, it is delayed by factors such as the travel distance and number of vehicles already on the road. The moving car keeps going through a loop of congested entries until its assigned exit attribute equals its final destination attribute. Once this loop sequence ends, the overall vehicle throughput and average travelling speed are then calculated by the simulation. The top-level model is equipped with a resolution factor variable. To prevent the system from growing beyond the software's transaction capacity, the resolution factor was set to 25. Thus, every moving entity within the system represents a group of 25 vehicles, travelling bumper-to-bumper. Additionally, a maximum batch number of entering vehicles at each entry ramp was set based on the total amount of people residing in the communities close to the corresponding ramp, and the pre-established evacuation ratio. This evacuation ratio is an estimate of the fraction of the population exiting from a particular area as determined by ALDOT officials. Other variables such as road length and number of lanes, average headway (i.e., the average distance between entities on the road), as well as lane occupancies are also used in the top-level model to determine the time a vehicle spends on the highway during the evacuation process, the travelled distances, evacuation rates, average delays, and travelling speeds. For example, the overall road occupancy level is increased according to the resolution factor and the rate of arrivals, and this leads to the calculation of the overall delay experienced by drivers already travelling on the Interstate.

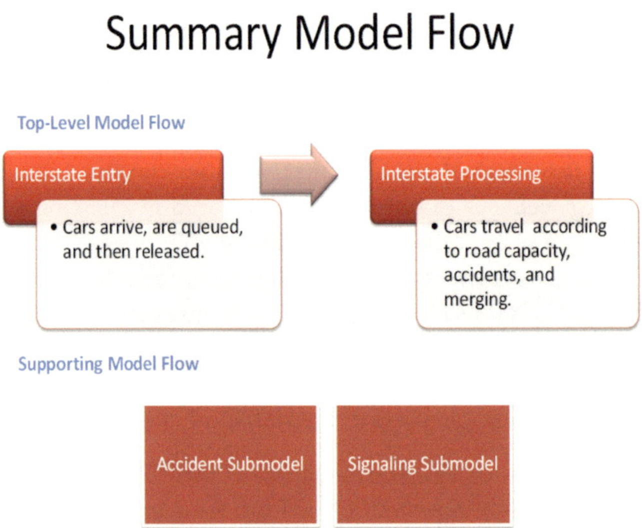

Figure 4. Evacuation simulation model architecture

The supporting models that contribute to the top-level model are the accident and signal models. The accident model determines the frequency of accidents on the highway during the evacuation process. Whenever the accident event is scheduled, a predetermined system delay becomes into effect in the model. The variables used in this supporting model (i.e. the accident factor and the accident delay time) are generated through user-defined probability distributions based on interviews with ALDOT transportation engineers. The other supporting model (the signal model) determines the timing of entities releases into the top-level model from the entry ramps. This is a stochastic process defined by pre-established probability distributions (i.e. Poisson and Binomial distributions) as well as heuristics established by the project analysts.

Figure 6 depicts the simulation logic for creating vehicles in the model. The traffic flow begins with the 20 ARRIVE blocks. Within each of these 20 ARRIVE blocks, the batch size is set equal to the quantity of entities that should be entering the interstate at each allowed time. This is accomplished by dividing the number of cars that will be allowed to enter at each exit by the resolution factor, and then, rounding the result to get the number of entities. "ANINT" is an internal system function which simply performs rounding [45].

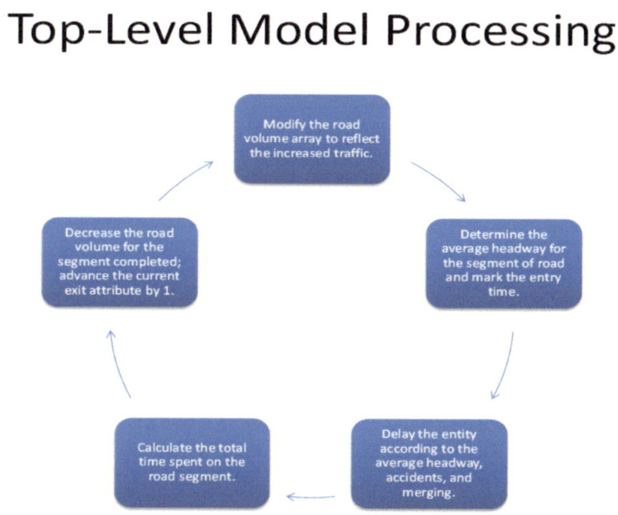

Figure 5. Transaction processing in the top-level model

The rate at which these batches are created is set equal to the frequency at which the cars are released. This is accomplished, as the next step in the ARRIVE blocks, by referencing the proper row in the SigDelay() array, which holds the three possible signaling frequencies to allow cars entry to the interstate. A second array, ExitSig(), stores the signal (1, 2, or 3) that each car will respond to at each exit. For example, the fourth ARRIVE block would have SigDelay(ExitSig(4)) as the formula for time between creations. This would reference the fourth row of the ExitSig() array, and discover that cars at the fourth exit are signaled by Signal 2. The formula would then reference the second row in the SigDelay() array to find the proper time between creations. The maximum number of batches, that each of these 20 ARRIVE blocks can create, is equal to the product of the total number of possible vehicles at the exit and the corresponding evacuation factor, divided by the resolution factor. This number is also rounded to be an integer

Within these ARRIVE blocks, the current time is indicated by the attribute TimeIn, and the entry exit index is indicated as ArrExit. The attribute CurrExit is set equal to ArrExit, and is later used to advance entities along the interstate. The final attribute set at this point is OffExit, which obtains its value from a discrete distribution with the help of the OffCumPr() array. Before the simulation begins, the user accesses the Off-Exit Cumulative Probability array, and inputs the cumulative probabilities that a car will get off the interstate at a specified exit number. OffExit references this array when drawing from the discrete distribution to find the entity's corresponding interstate exit. If, by chance, OffExit happens to be less than or equal to ArrExit, OffExit will be set to 20, that is, the last exit of the Interstate. Entities then proceed to the WAIT block, where they are held until the proper signal for their exit occurs. The proper signal for the exit is found by referencing the ExitSig() array. For example, the 14th ARRIVE block, and thus the 14th WAIT block, would find its signal through the expression ExitSig(14). The release limit is set equal to the number of cars that will be released at each signal divided by the resolution factor.

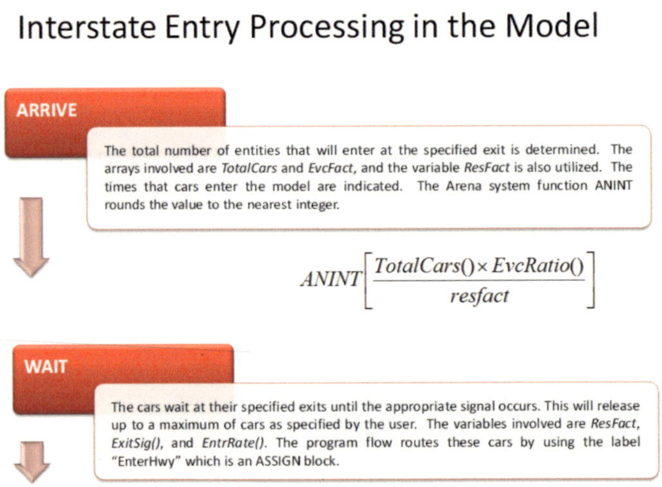

Figure 6. ARENA processing of vehicles entering the interstate roadway

When an entity (i.e. vehicle) enters the interstate (see Figure 7), it goes to the first ASSIGN block on the upper left. First, the road volume array RoadVol(), which stores the total number of cars is adjusted upward by the resolution factor. Since the entity's current exit is stored in the attribute CurrExit, the road volume can be adjusted upward simply by the formula RoadVol(CurrExit) = RoadVol(CurrExit) + resfact. After the road volume array compensates for the additional cars, the attribute AvHeadwy, which represents average headway, is calculated. Basically, this calculation takes the length of the road, multiplies it by the number of lanes on the road, and subtracts out the length of road that all the cars are using (assuming an average car length is 15 ft). All of this is then divided by the length of road that the cars are using to ultimately come out with the average headway. The constant 5,280 represents a unit conversion from miles to feet, and the constant 15 represents the average car length. Finally, the time in to the current exit, ExTimeIn, is marked so the total time spent on the current strip of road can be calculated.

Also, as noted in Figure 7, the time to be spent on the current road segment is calculated in the DELAY block. The formula references a table referred to as SpeedTbl, which uses the assumption that a linear relationship exists between average headway and speed (1 car length = 10 mph), and that the maximum speed on the interstate is 70 mph. Table functions in Arena will automatically interpolate, so when an average headway of 2.37 ft/car is used as an input, the function will output 23.7 mph. In the delay formula, the constant of 60 is used for unit conversion purposes, i.e. that 1 hour = 60 minutes. AccFact() and MrgFact() are arrays that contain values between 0 and 1 which will delay the traffic proportionally according to the factors they present. For instance, if the 16th exit had an accident, AccFact(16) would be set to 0.6 to reflect that the accident is slowing traffic down to 60% of what it would be otherwise. Additionally, if the ninth exit is left open for cars to enter the Interstate, MrgFact(9) would likely be set equal to a number less than 1, whereas closed exits would maintain a factor of 1 since merging would not be affecting traffic at closed exits. "TF" is a system function of Arena that stands for "Table Function." Unit analysis shows that the final result is in minutes, since the numerator is (miles * minutes/hour) and the denominator is (miles/hour).

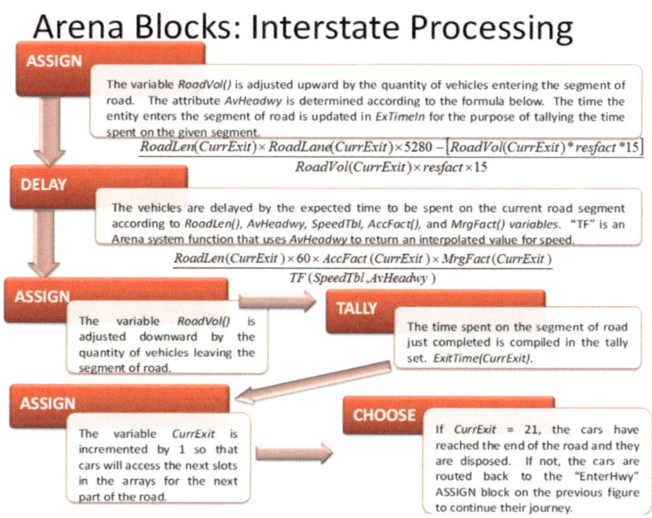

Figure 7. ARENA logic for vehicles on the interstate roadway

The next ASSIGN block, in Figure 7, adjusts the road volume downward as the entity has now completed the current road segment. The TALLY block subsequently records the total time spent on the specific exit. The next ASSIGN block advances CurrExit by 1 as the entity has completed the road segment, and is now advancing to the next segment. Finally, the CHOOSE block checks to see whether the entity has reached the exit it will leave the interstate (if CurrExit = OffExit) or to see if the car has reached the Montgomery contraflow terminus (if CurrExit = 21). If either of those conditions is true, the entity is disposed. Otherwise, the entity is routed back around to the first ASSIGN block in Figure 7 where the next segment of road has its road volume incremented upward.

Dummy transactions are used in the simulation model to represent accidents, and are created at a set time interval (in minutes) stored in a single variable called *AccFreq*. For each simulated accident event, an *ASSIGN* block allocates the exit where the accident is to occur

(*AccExit*) according to a user-defined probability distribution named *AccCumPr()*, as well as a factor to represent the resulting slower traffic (*AccFact*). Under ideal conditions, *AccFact* is equal to 1 for any given road segment. However, after the *ASSIGN* block determines that an accident has happened at a particular exit, the respective *AccFact* is re-set to a number smaller than 1 (e.g., 0.7) based on the probabilistic equations embedded in the model. The accident is allowed to persist for a certain duration of time, *AccRcvr*. After that time has elapsed, the accident factor for that specific exit is returned to one. Figure 8 depicts the ARENA logic embedded in the simulation's Accident Submodel.

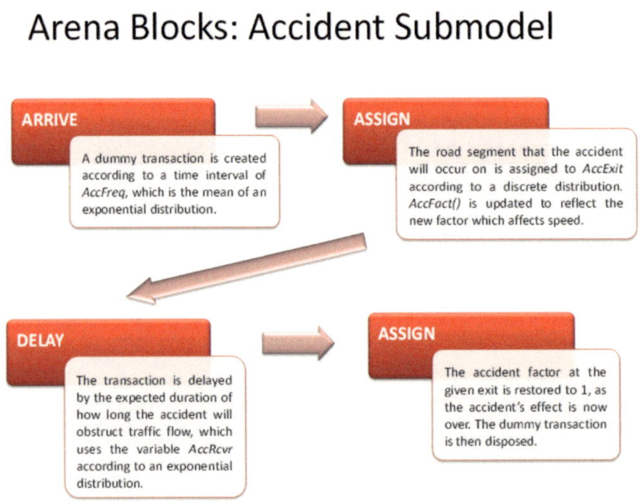

Figure 8. ARENA logic for considering the effects of accidents on the traffic flow

7. Results

With parameters set for the initial model, i.e. all 20 ramps open, the simulation was run for a twelve hour period, consistent with the duration of one of the previous I-65 contraflow operations. The resolution factor, i.e., the entity to car ratio, was set to 25, and the entry rates for each on-ramp were set in accordance with the data presented in Tables 1 and 2. For this initial model, a greater number of vehicles were allowed to get on the Interstate closer to the coast than at other entry points located further north, representing actual observed evacuation behavior [44]. Hence, such on-ramps experienced a faster rate of signals for the release of batches of vehicles. Due to the limited availability of hotel accommodations for evacuees, between Mobile and Montgomery, the final destination attribute for each simulated vehicle was established through a probability scheme based on the assumption that there is an equal chance (i.e. 20%) for a travelling car to exit the interstate at any of the five exit ramps of I-65 in the City of Montgomery. Test statistics were collected from the simulation after 30 independent replications. The average number of vehicles that reached their final destination during any 12-hour period was 113,318. That is, with all on-ramps open for traffic evacuation, the average evacuation rate was 9,442 vehicles/hour.

Alternative Utilizing Closure of Exits 130 and 57:

IDENTIFIER	ESTD. MEAN DIFFERENCE	STANDARD DEVIATION	0.950 C.I. HALF-WIDTH	MINIMUM VALUE	MAXIMUM VALUE	QTY OBS
Cars	135	136	273	4.01e+003	6.88e+003	30
				3.87e+003	5.55e+003	30

FAIL TO REJECT H0 ="/ MEANS ARE EQUAL AT 0.05 LEVEL

Alternative Utilizing Closure of Exit 142:

IDENTIFIER	ESTD. MEAN DIFFERENCE	STANDARD DEVIATION	0.950 C.I. HALF-WIDTH	MINIMUM VALUE	MAXIMUM VALUE	QTY OBS
Cars	-13.6	131	262	4.01e+003	6.88e+003	30
				3.97e+003	5.57e+003	30

FAIL TO REJECT H0 ="/ MEANS ARE EQUAL AT 0.05 LEVEL

Alternative Utilizing Closure of Exit 128:

IDENTIFIER	ESTD. MEAN DIFFERENCE	STANDARD DEVIATION	0.950 C.I. HALF-WIDTH	MINIMUM VALUE	MAXIMUM VALUE	QTY OBS
Cars	-39.7	145	290	4.01e+003	6.88e+003	30
				3.92e+003	6.41e+003	30

FAIL TO REJECT H0 ="/ MEANS ARE EQUAL AT 0.05 LEVEL

Alternative Utilizing Closure of Exit 54:

IDENTIFIER	ESTD. MEAN DIFFERENCE	STANDARD DEVIATION	0.950 C.I. HALF-WIDTH	MINIMUM VALUE	MAXIMUM VALUE	QTY OBS
Cars	56.1	138	276	4.01e+003	6.88e+003	30
				3.79e+003	5.46e+003	30

FAIL TO REJECT H0 => MEANS ARE EQUAL AT 0.05 LEVEL

Table 3. Statistical analysis using two sample t means comparisons

The main focus on the study was to investigate that effect that selectively closing entry ramps from entering traffic had on the overall evacuation process of the Alabama Gulf region during a hurricane situation. The closing of entry ramps was modeled by re-directing vehicles from the arrival at the closed on-ramp to subsequent entry points further north, taking into consideration the time of travel through side roads, the availability of these side roads, as well as the open entry ramps' own traffic volumes. In the simulation, this traffic rerouting was conducted through a random variable distribution of vehicles to entry ramps located no further than thirty miles away from the closed entry point. This was a combinatorial procedure consisting of two phases. The first step was the analysis of twenty simulation models each pertaining to the closing of a particular entry ramp in the model. The next step was to simulate the closing of multiple entry ramps through permutations of two and three entry ramps closing at a time.

To guarantee statistical soundness of the hypothesis testing procedure, all simulation trials involved 12 hours of simulated time, each replicated 30 times. The null hypothesis of the study implied that the means for the evacuation rate remained the same for the original model with all 20 entry ramps open, and for every other permuted model under examination. The hypothesis testing was performed in a pair-wise fashion, i.e. original model versus alternate model, and an alpha level of 5% was used in all tests. Statistical testing was performed on the data generated by the simulation model to identify variation in relevant traffic variables affecting the north-bound traffic flow. According to Braess' Paradox [36], having all ramps open to vehicles exiting the region may reduce the evacuation traffic flow since in-flow congestion may be generated by entry ramps located within a few miles from each other; and due to the difficulty encountered by emergency and law-enforcement vehicles bound South when all ramps are exclusively for north-bound traffic.

Table 3 displays the most pertinent results from the statistical analyses. The authors determined that there are five entry ramps that can be strategically controlled by ALDOT personnel during an evacuation situation without slowing down the overall evacuation traffic flow. These entry ramps are Exit 54, Exit 57, Exit 128, Exit 130, and Exit 142. Exits 54, 128, and 142 can be closed from entering traffic individually without affecting of overall evacuation rate of vehicles. When closed individually, exits 57 and 130 posed a detrimental effect on the overall evacuation effort. However, it is interesting to note that when they are closed at the same time, their combined effect does not alter the overall flow of evacuating traffic on the interstate.

8. Further research

It is proposed to begin investigation of the larger roadway network through a further evolution of this research to better meet the needs of emergency planners. Decision support systems (DSS) are software systems that utilize sophisticated algorithmic approaches to address problems. Within a DSS, the model base contains the specific analytical methods used for processing the accessed data. The utilization of computer-based simulation, within a DSS, is a long-accepted means of modeling complex systems and activities. The objective of this proposed effort is to establish an incident evacuation decision support system employing a series of traffic analysis algorithms that consider: 1) identification of prevalent traffic flow conditions during a predetermined time window, 2) recognition of incident occurrence, 3) incident characterization, and 4) subsequent routing. These routing algorithms would provide the basis for a network simulation model of the three-state region (Alabama, Louisiana and Mississippi). By employing actual or simulated traffic sensor input, planning alternatives can be evaluated with consideration of traffic congestion levels and adverse roadway conditions. The integration of DSS capabilities within a traffic data collection system framework will provide a means to identify high-leverage areas where manpower resources could be applied in order to improve the overall traffic flow and throughput of the evacuation route. This will allow the envisioned system to act as an intelligent filter and highlight current problems so that emergency personnel can quickly address them. By conducting what-if analysis, the enhanced system can project the effects of changes in selected variables on the overall traffic flow. This aspect is vital in case of hurricane evacuation activities, when the path of a hurricane quickly shifts, requiring rapid replanning. The various decision alternatives can then be evaluated economically without disrupting existing operations, or incurring unnecessary costs.

9. Conclusions

Vehicle usage of roadways continues to increase across the United States, and results in many roadways operating at near capacity during normal peak periods [46]. This situation is compounded under emergency evacuation situations. Effective planning can decrease traffic congestion, fuel consumption, and the response time of emergency vehicles. It has, in recent years, taken on added significance for federal, state, and local governments by reducing delays and increasing the number of vehicles evacuated from hazardous areas [47]. A variety of algorithmic methods, ranging from simple statistical tools to probabilistic neural networks, have been successfully developed to aid traffic planners. A number of them have been applied to identifying areas of potential traffic congestion under an evacuation situation. Fonseca et al. [48] provide a survey of the literature in this area. The overall traffic capacity of the affected areas can be increased by strategically selecting evacuation measures oriented to the avoidance of congestion. This paper discusses the incorporation of Braess' Paradox within a computer-based simulation framework to better evaluate evacuation traffic throughput.

Alabama traffic officials and emergency personnel want to have the most effective way of ensuring the safety of coastal residents when the danger of severe tropical weather is eminent. Through this study, a comprehensive simulation of I-65, within the State of Alabama, was conducted to access its effectiveness as potential evacuation route during a hurricane situation. It was discovered that by having all 20 entry ramp exits open to merging vehicles bound north, an average of 113,318 vehicles will reach the City of Montgomery safely within a 12-hour period. However, having all exits blocked from traffic bound south poses great difficulty to emergency-response officials needing to access areas in the path of the storm. Through a rigorous and exhaustive process, it was determined that five of the twenty ramps can be strategically controlled to resolve conflicts when restricted flow of vehicles bound south on the Interstate shoulder is needed (as in the case of emergency vehicles responding to accidents on the reverse-laned I-65). The authors have developed a detailed simulation model to analyze discrete networks representing the flow of traffic along planned evacuation routes, with consideration of the effects of Braess' Paradox. The findings of this study serve as evidence for the need of similar studies to be conducted in other main routes of contraflow evacuation along the coastal areas of the United States. Although other areas in the U.S., as well as other countries, may not utilize contraflow, the application of simulation modeling has proven to be an effective tool for hurricane evacuation planning.

Acknowledgements

Special thanks are extended to Jordan Johnston and Chase Jennings for their invaluable help in gathering data for this project. The authors also wish to thank Qingbin Cui for his assistance.

Author details

Gary P. Moynihan[1] and Daniel J. Fonseca[2]

*Address all correspondence to: gmoynihan@eng.ua.edu

1 Department of Civil, Construction and Environmental Engineering The University of Alabama, Tuscaloosa, AL, USA

2 Department of Mechanical Engineering The University of Alabama, Tuscaloosa, AL, USA

References

[1] Blake, E., Rappaport, E., & Landsea, C. (2007). The Deadliest, Costliest, and Most Intense. *United States Tropical Cyclones from 1851 to 2006 (and Other Frequently Requested Hurricane Facts), National Oceanic and Atmospheric Administration Technical Memorandum NWS TPC-5.*

[2] Texas State Government. (2012). *Texas.gov Emergency Portal Page,* Available at:, http://emergency.portal.texas.gov/en/Pages/Home.aspx,, Accessed 2012 April 27.

[3] Texas Department of Transportation. (2012). *Hurricane Evacuation Contraflow Evacuations,* Available at:, http://www.dot.state.tx.us/travel/contraflow_publications.htm,, Accessed 2012 April 27.

[4] Hibbard, J., & Hodges, D. (2005). *Technical Memorandum: Contraflow Plans for the Florida Interstate Highway System,* Available at:, http://www.floridaits.com/PDFs/TWO45ContraFlow/050606-FIHS Contraflow Rprt-1pdf, Accessed 2012 April 27.

[5] Mississippi Department of Transportation. (2009). *Contraflow Plan for Interstate Hurricane Evacuation Control,* Available at:, http://www.gomdot.com/home/EmergencyPreparedness/pdf/ContraflowPlan.pdf,, Accessed 2012 April 27.

[6] National Weather Service,. (2012). *National Hurricane Center Webpage,* Available at:, http://www.nhc.noaa.gov,, Accessed 2012 April 18.

[7] Gerdes, B. (2007). *Percent of Alabamians Live in Hurricane Counties, Accordin g to UA's State Data Center, NOAA, The University of Alabama News,* Available at:, http://uanews.ua.edu/anews2007/jun07/counties061307.htm,, Accessed 2012 April 27.

[8] Alabama Department of Tr an sportation (2008). Plan for Reverse-Laning Interstate I-65 in Alabama for Hurricane Evacuation Available at: http://www.dot.state.al.us/maweb/reverse_laning_plan_summary.htm, Accessed April 12, 2012.

[9] Wolshon, B. (2001). One-Way-Out: Contraflow Freeway Operation for Hurricane Evacuation. *Natural Hazards Review,* 2, 105-112.

[10] Balakrishna, R., Wen, Y., Ben-Akiva, M., & Antoniou, C. (2008). Simulation-Based Framework for Transportation Network Manag ement for Emergencies, Transportation Research Record:. *Journal of Transportation Research Board,* 2041, 80-88.

[11] Li u, Y., Chang, G., Liu, Y., & Lai, X. (2008). Corridor-Based Emergency Evacuation System for Washington, D.C.: System Development and Case Study, Transportation Research Record:. *Journal of Transportation Research Board,* 2041, 58-67.

[12] Brown, C ., White , W., van Slyke, C., & Benson, J. (2009). Development of a Strategic Hurricane Evacuation- Dynamic Traffic Assignment Model for the Houston, Texas, Region, Transporta tion Research Record:. *Journal of Transportation Research Board,* 2137, 46-53.

[13] Sbayti, H., & Mahmassani, H. (2006). Optimal Scheduling of Evacuation Operations, Transportation Research Record:. *Journal of Tra nsportation Research Board,* 1964, 238-246.

[14] Chiu, Y., Zheng, H., Villalobos, J., & Gautam, B. (2007). Modeling No-noti ce Mass Evacuation Using Dynamic Traffic Flow Optimization Model,. *IIE Transactions,* 39, 83-94.

[15] Pel, A., & Bliemer, M. (2008). Evacuation Plan Evalua tion: Assessment of Mandatory and Voluntary Vehicular Evacuation Schemes by means of an Analytical Dynamic Traffic Model, Compendium of Papers DVD, The 87th Transportation. *Research Board Annual Meeting, Washington, D.C., January 2008m,* 08-2086.

[16] Yao, T., M andala, S., & Chung, B. (2009). Evacuation Transportation Planning under Uncertainty: A Robust Optimization Approach,. *Network Spatial and Economics,* 9, 171-189.

[17] Ng, M., & Waller, T. (2010). Reliable Evacuation Planning via Demand Inflation and Supply Deflation. *Transportation Research, Part E,* 46, 1086-1094.

[18] Kim, S., Shekhar, S., & Min, M. (2008). Contraflow Network Reconfiguration for Evacuation Route Planning,. *IEEE Transactions on Knowledge and Data Engineering,* 20, 1115-1129.

[19] Lv, N., Yan, X., Xu, K., & Wu, C. (2010). *Bi-level Programming based Contraflow Optimization for Evacuation Events, Kybernetes*, 39, 1227-1234.

[20] Theodoulou, G., & Wolshon, B. (2004). Alternative Methods to Increase the Effectiveness of Freeway Contraflow Evacuation, Transportati on Research Record:. *Journal of Transportation Research Board*, 1865, 48-56.

[21] Williams, B., Tagliaferri, A., Meinhold, S., Hummer, J., & Rouphail, N. (2007). Simulation and Analysis of Freeway Lane Reversal for Coast al Hurricane Evacuation,. *ASCE Journal of Urban Planning and Development*, 133, 61-72.

[22] Meng, Q., & Khoo, H. (2008). Optimizing Contraflow Scheduling Problem: Model and Algorit hm,. *Journal of Intelligent Transportation Systems*, 12, 126-138.

[23] Pal, A., Triche, M., Graettinger, A., Rao, K., Mc Fadden, J., & Turner, D. (2005). Enhancements to Emergency Evacuation Procedures,. *Final Research Report. University of Alabama Transportation Center.*

[24] Sisiopiku, V. (2007). Development of Dynamic Traf fic Assignment Model to Evaluate Lane Reversal Plans for I-65,. *University of Alabama- Birmingham School of Engineering,* Available at, http://main.uab.edu/soeng/TemplatesInner.aspx?pid=98705, A ccessed April 11, 2012.

[25] Turban, E., & Aronson, J. (2001). Decision Support Systems and Intelligent Systems. *Upper Saddle River, NJ: Prentice Hall.*

[26] Hillier, F., & Lieberman, G. (2005). *Introduction to Operations Research (8th Edition)*, McGraw-Hill, New York.

[27] Le Blanc, L. (1975). An Algorithm for the Discrete Network Design Problem. *Transportation Science*, 9, 183-199.

[28] Magnanti, T. L., & Wong, R. T. (1984). Network Design and Transportation Planning: Models and Algorithms,. *Transportation Science*, 18, 1-55.

[29] Yang, Hai., & Bell, G. H. M. (1998). Models and Algorithms for Road Network Design: A Review and Some New Developments. *Transportation Reviews*, 45-58.

[30] Solanki, S., Rajendra, Gorti. K., & Jyothi, Southworth. F. (1998). Using Decomposition in Large-Scale Highway Network Design with A Quasi-Optimization Heuristic,. *Transportation Research B*, 32, 127-140.

[31] Wardrop, J. (1952). Some Theoretical Aspects of Road Traffic Flow Research,. *Proceeding of the Institute of Civil Engineesr II*, 1, 325-378.

[32] Steinberg, R., & Zangwill, W. (1983). The Prevalence of Braess' Paradox,. *Transportation Science*, 17, 301-318.

[33] Tung, S. (1986). *Designing Optimal Networks: A Knowledge-Based Computer Aided Multicriteria Approach", Ph.D. Dissertation.*, University of Washington, Seattle, WA.

[34] Friesz, T. L., & Shah, S. (2001). An Overview of Nontraditional Formulations of Static and Dynamic Equilibrium Network Design,. Transportation Research, Part B. , 35, 5-21.

[35] Murchland, J. D. (1970). Braess' Paradox of Traffic Flow,. *Transportation Research*, 12, 391-394.

[36] Cohen, J., & Kelly, F. (1990). A Paradox of Congestion in a Queuing Network,. *Journal of Applied Probability*, 27, 730-734.

[37] Bass, T. (1992). Road to Ruin,. *Discover*, 3, 56-61.

[38] Fonseca, D. J., Daosuparoch, S., Moynihan, G. P., & Chen, D. S. (2003). A Computer-Based System for Road Selection,. *Expert Systems*, 20, 133-140.

[39] Fu, H., & Wilmot, C. (2004). *A Sequential Logit Dynamic Travel Demand Model, Proceedings of the TRB Annual Meeting*, Available at:, http://www.ltrc.lsu.edu/pdf/TRB2004000960.pdf,, Accessed 2012 April 14.

[40] Nasser, M., & Birst, S. (2010). *Mesoscopic Evacuation Modeling for Small- to Medium-Sized Metropolitan Areas, Report to the Advanced Traffic Analysis Center, Upper Great Plains Transportation Institute*, North Dakota State University, Fargo, ND, Available at:, http://www.mountain-plains.org/pubs/pdf/MPC10222.pdf, Accessed 2012 April 14.

[41] Tanaka, K. (2007). Traffic Congestions and Dispersion in Hurricane Evacuation,. *Physica A: Statistical Mechanics and Its Applications*, 376, 617-627.

[42] U.S. Army Corps of Engineers . (2001). *Alabama Hurricane Evacuation Study- Summary Report*, Available at:, http://chps.sam.usace.army.mil/USHESdata/Alabama/ALmain-report.htm,, Accessed 2012 April13.

[43] National Oceanic and Atmospheric Administration. (2011). *Hurricane Planning and Evacuation Assessment Reports*, Available at:, http://www.csc.noaa.gov/hes/hes.html, Accessed 2012 April 13.

[44] U.S. Army Corps of Engineers . (2001). *Alabama Hurricane Evacuation Study- Transportation Study*, Available at:, http://chps.sam.usace.army.mil/USHESDATA/Alabama/altranspage.htm, Accessed 2012 April 12.

[45] Kelton, W. D., Sadowski, R. P., & Sturrock, D. T. (2001). *Simulation with Arena (4th ed.)*, New York, Prentice Hall.

[46] Nowakowski, C., Green, P., & Kojima, M. (1999). Human Factors in traffic Management Centers:. A Literature Review (Technical Report UMTRI-99-5) The University of Michigan Transportation Research Institute, Ann Arbor, MI.

[47] Lomax, T. J., Turner, S. M., & Margiotta, R. (2003). Monitoring Urban Roadways in 2001: Examining Reliability and Mobility with Archived Data (Report FHWA-OP-03-041),. *United States Federal Highway Administration, Washington, D.C.*

[48] Fonseca, D. J., Moynihan, G. P., & Fernandes, H. (2011). The Role of Non-Recurring Congestion in Massive Hurricane Evacuation Events, in Recent Hurricane Research:. *Climate Dynamics and Societal Impacts (A. Lupo, ed.),* InTech Publishing, Rijeka, Croatia, 441-458.

Chapter 7

Transport of Nitrate and Ammonium During Tropical Storm and Hurricane Induced Stream Flow Events from a Southeastern USA Coastal Plain In-Stream Wetland - 1997 to 1999

J. M. Novak, A. A. Szogi, K.C. Stone, X. Chu,
D. W. Watts and M. H. Johnson

Additional information is available at the end of the chapter

http://dx.doi.org/10.5772/51152

1. Introduction

Nitrogen (N) is an essential element for the functioning of aquatic ecosystems [1, 2]; however, when in excess, it can be detrimental to water quality by promoting eutrophication [3, 4] and growth of harmful microalgae and dinoflagellates [5]. Elevated N levels transported by streams and rivers into downstream coastal estuaries are of particular concern because eutrophication can affect up to 65% of the estuarine area of the coastal USA [6]. Eutrophication leads to hypoxia which stresses fish, shellfish, and invertebrates, and long-term exposure to hypoxia is fatal to most endemic fauna [7, 8]. Rapidly growing and diversifying non-point and point sources of N (e.g. agricultural crop production, urban wastewater, fertilizer use, N-enriched rainfall, and concentration of animal production systems) have been linked as causes for these troubling symptoms of eutrophication [9-12].

In eastern North Carolina, the Cape Fear River has experienced water quality issues due to high N input loads from non-point sources [13]. Mallin et al. [13] has linked water quality degradation in the Cape Fear River to N-enriched runoff from land areas with highly concentrated animal production. North Carolina has about 10 million pigs with the majority (58% of state total) of production located within the Cape Fear River basin in four counties (Duplin,

Sampson, Bladen, and Wayne) [14]. The Cape Fear and Lower Cape Fear Rivers flow through or receive input from small tributaries that drain watersheds within these four North Carolina counties. Nitrogen transported from small tributaries into the Cape Fear and Lower Cape Fear River systems has caused concern about water quality degradation [12, 15].

Non-point N movement into river and streams is a serious water quality issue in watersheds of the southeastern Coastal Plain region [16]. Nitrogen stored in soils or groundwater can move in runoff and in subsoil flow into rivers and streams especially after storm events [15] resulting in increased nutrient loading into downstream estuaries [13]. Because stream and rivers can readily transport N into estuaries, understanding processes that influence N cycling along the flow continuum offers the opportunity to shape best management strategies to reduce excess N movement. Galloway et al. [17] estimated that almost 50% of the N entering streams and rivers can be removed by N assimilative processes before it reaches coastal waters. Assimilative N processes in wetlands involves reactions between the water column and sediments [18, 19], and by microbial assimilation, denitrification and plant uptake [20, 21].

Wetlands are natural landscape features that provide an ecosystem service for N removal and are effective at reducing N loads [22, 23] and concentrations [24]. Hunt et al. [23] showed that a North Carolina Coastal Plain in-stream wetland annually removed 3 kg N ha^{-1} d^{-1} or about 37% of the total N as inflow. However, further examination of stream N export loads during direct runoff and base flow events as well as shifts between N storage pools within the wetland is needed to better understand N dynamics in this Coastal Plain ISW. It was hypothesized that greater N loads should be exported during periods of high stream outflows due to shifts in the wetland's N storage pools. This is a plausible premise because precipitation from storm events should increase stream inflows thus causing hydrologic disturbances within the wetland. In turn, these hydrologic disturbances stir up N associated with sediments, shifting equilibrium reactions between nutrients in the water column and sediment with increasing N losses with outflow. Wetland N storage/release dynamics are germane for this agriculturally intensive region of North Carolina because of the frequency of tropical storms and hurricane events [25, 26] and the subsequent increase in N movement into coastal estuaries [9, 13, 5].

The data of this study was extracted from a decades-long data base gathered from a US Department of Agricultural Water Quality Demonstration Project. The 3-year hydrological data presented in this article contained significant tropical storm and hurricane activity that is still relevant today. Current hurricane activity and tropical storms projections suggest an increase in their activity due to climate variability. Thus causing a need for past data contributions for future model projections. The objectives were to *i*) estimate annual Q_t, Q_{dr} and Q_b, and annual N (as NO_3-N and NH_4-N) and calculate loads exported from the ISW during these events; and *ii*) ascertain annual shifts in N storage pools between N associated with sediment, the pore water and water column.

2. Materials and methods

2.1. Watershed and In-stream Wetland Description

This study was carried out in the Herrings Marsh Run (HMR) watershed of Duplin Co., North Carolina. The HMR watershed is located in the Middle Coastal Plain region and is underlain with sandy to clay-enriched marine sediments [27]. The landscape topography is typical of the southeastern USA Coastal Plain possessing wide, nearly level to gently sloping upland areas, which have been dissected by primary, secondary and tertiary streams [27]. Soils that form in the upland areas are sandy with internal drainage ranging from somewhat poorly to excessively well-drained. Fairly wide (3-to 15-m) riparian zones form along streams resulting in floodplain areas containing very poorly drained soils [28].

For this study, daily precipitation totals were collected from both Warsaw and Clinton, North Carolina [29]. Technical information was gathered on characteristics of tropical storms and hurricanes (dates making North Carolina coastal contact, reported dates over Duplin Co., and daily precipitation) in reference [29].

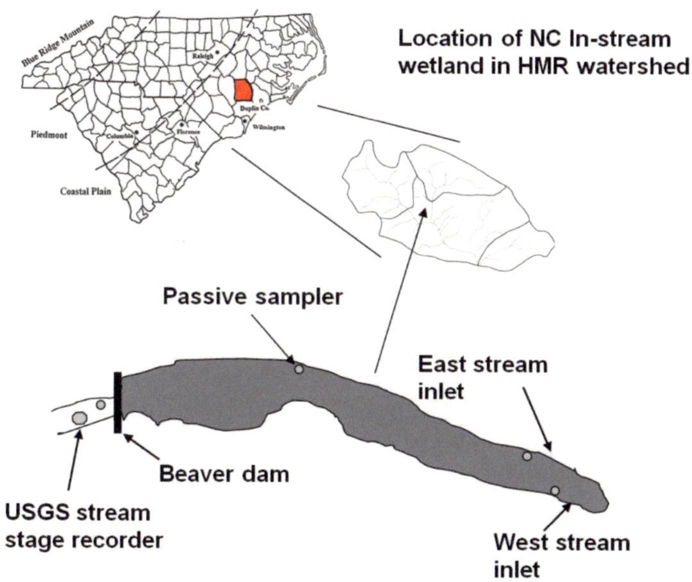

Figure 1. Location of the Herrings Marsh Run watershed in Duplin Country, North Carolina. The in-stream wetland was located at the outlet of a subwatershed and equipped with an outflow stage recorder with passive samplers installed at 4 different locations (outlet, mid-point, and two inlets).

Crop and animal production practices in the HMR watershed are typical for southeastern North Carolina [28]. This includes row and truck crop production and different livestock operations. Based on a 1993 animal survey of the HMR watershed, animal heads were estimated at 29,931 for hog, 94,000 for poultry, and 176 for cattle [28]. Liquid animal manure from these operations is typically applied to row crops and pastures as a source of fertilizer.

In-stream wetlands in the HMR watershed are commonly formed in flat topographic locations and in shallow depressions along the stream continuum [27]. The studied wetland was located in the north central part of the watershed and is about 3.3 ha in size, has two diffuse stream inlets, storage volume of 29,000 m^3, and the pond depth varied from < 0.3 m to almost 2 m deep (Figure 1). It receives inflowing water from two shallow second- and third-ordered black water streams on its eastern and western sides (Figure 1). This ISW is bordered by deciduous trees consisting of Bald Cyprus [*Taxodium distchum* L.], Swamp Chestnut Oak [*Quercus michauxii* Nutt.], Red Maple [*Acer rubrum*, L.], Green Ash [*Faxinus pennsylvanica* Marsh.], and both Loblobby [*Pinus taeda*, L.] and Longleaf [*Pinus palustris*, L.] pines. These trees offered a food source for colonization of the ISW by beavers (*Castor Canadensis*). Consequently, over several years, the beavers dammed up the outlet (Figure 2), which caused the two stream inflow locations to flood frequently.

2.2. Stream Sampling Locations and Stream Flow Estimates

In the early 1990s, both the east and west stream inlets flowed in well-defined stream channels. To measure water inflows, H-flumes were installed at these two inlets (Free Flow, Omaha, Nebraska†) equipped with pressure transducers [23]. Meanwhile, beavers improved the dam that slowed outflow and increased flooding which swamped both H- flumes. The flooding incidences lasted over several months causing inaccurate stream inflow measurements. So, there are no true inflow Q measurements available for this study.

Outflow discharges from the ISW were measured as outlined in [23] using a U.S. Geological Survey automated gauging station. Estimates of mean Q_t (as $m^3\ d^{-1}$) were obtained electronically from Q_t results that were recorded at 15-min intervals using an automated water level recorder. There were periods when daily Q_t was not recorded due to equipment damage by storms and flooding, so a linear interpolation method was used to compute values for these missing periods. The daily Q_t results were further separated into Q_{dr} and Q_b using the digital filter method [30, 31]. This digital filtering method served to partition the total flow using equations 1 to 2:

$$Q_{dr,k} = \alpha Q_{dr,k-1} + 0.5(1 + \alpha)(Q_{tk} - Q_{tk-1}) \quad (1)$$

$$Q_{b,k} = Q_{tk} - Q_{dr,k} = 0.5(Q_{tk} + Q_{tk-1}) - 0.5\alpha(Q_{tk} - Q_{tk-1}) - \alpha Q_{dr,k-1} \quad (2)$$

Where $Q_{dr,k}$ = direct runoff at time step k; $Q_{b,k}$ = base flow at time step k; Q_{tk} = total stream flow at time step k; and α = filter parameter. A α of 0.925 was used in the iterations [31]. Although there are method and assumption biases in this flow separation method, Nahatan and McMahon [31] recommend the use of the digital filter method when daily Q_t results are available. Stream outflow samples were collected every 2 h using American Sigma automatic water samplers (Danaher Corp., Loveland, Colorado) and later combined to make composite 3.5-d samples.

Figure 2. In-stream wetland outlet showing remnants of beaver dam (top), installation of passive samplers (bottom left), and removal of water sample using syringe from cells in passive sampler (bottom right).

2.2.1. Dissolved N Measurements

Sample preservation was accomplished by adding dilute H_2SO_4 to each automated water sampler bottle before sample collection. The acidified samples were collected weekly, filtered (0.45 μm), and analyzed for NH_4-N and NO_3-N using a TRAACS 800 Auto-Analyzer (Bran + Luebbe, Buffalo Grove, Illinois) with EPA methods 353.2 and 350.1, respectively [32]. Quality control protocols for these analyses were outlined in reference [23].

2.2.2. Dissolved N Mass Loads in Wetland Outflow

From January 1997 to October 1999, daily outflow NH_4-N and NO_3-N mass loads (kg d^{-1}) were calculated by multiplying the daily mean flows with NH_4-N and NO_3-N by using their interpolated concentrations over the 3-to 4-d interval [33]. In this data set, the missing stream N concentrations were linearly interpolated to provide their daily estimated concentrations, then were used to compute corresponding daily N loads.

2.3. Passive Samplers Installation, Sediment and Water Sample Collection

Plexiglass passive samplers (peepers) were used to sample sediment pore water and the overlying water column according to procedures outlined in references [26, 33]. Each peeper consisted of a long block of plexiglass into which a series of cells spaced at 1-cm-depth increments were filled with deionized water and sealed with a membrane. Pairs of passive samplers were placed about 12-14 cm deep into the sediment at 4 locations (Figures 1 and 2) during a two-week period in Aug./Sep. during 1997, 1998 and 1999. Placement of the peepers at both ISW inlets were in shallow (0.2 to 0.4 m deep) areas, while placement at the midpoint and outlet sites was in deeper sections (0.5 to 0.8 m deep). Peepers were allowed to equilibrate with the sediment pore water and water column for 2 weeks. After removal of the peepers from the sediment, a plastic syringe was used to withdraw the sample from each cell in the peeper. The liquid sample was transferred into a plastic bottle containing dilute H_2SO_4 (preservative). The liquid samples were analyzed for NH_4-N and NO_3-N using the colorimetric method described earlier. Exchanges between water column and sediment N pools were determined by comparing relative differences in the water column and sediment pore water NH_4-N and NO_3-N concentrations. Sediment cores were also collected within 1 m of each peeper site using a bucket auger to 20 cm deep after peepers were installed. The sediments were air-dried and ground to pass a 2-mm sieve. The TKN concentrations were measured in all sediments using EPA method 351.2 [32]. Changes in the ISW N pools were determined by comparing relative differences in the annual sediment TKN and sediment pore water NO_3-N and NH_4-N concentrations.

2.4. Statistics

Outflow Q results, N concentrations and loads were initially computed on a daily basis and were also reported on a monthly and annual basis. This allowed for estimates of the cumulative NO_3-N and NH_4-N masses exported over the three-year study. To determine if the annual N loads (expressed as kg ha^{-1}) were dependent on outflow characteristics, the compiled mass of NO_3-N and NH_4-N by year were linearly regressed against the annual Q_t, estimates. This computation was accomplished by dividing the annual N loads by the subwatershed area (425-ha). All statistical analyses were performed using SigmaStat software version 3.5 (SSPS, Chicago, Illinois).

3. Results and Discussion

3.1. Storm Activity

In the Atlantic Ocean, a distinct tropical storm (TS)/hurricane season occurs from June 1st to November 30th, with more frequent storm activity reported in August to September [34]. In this study, TS and hurricane activity was also active in August through September (Table 1). Tropical storm 1 in 1997 deposited 70 mm of precipitation over the watershed. This storm prompted an increase in Q_t from the ISW from 872 m^3 d^{-1} to 9.7 x 10^3 m^3 d^{-1} within 24 h of the

http://dx.doi.org/10.5772/51152

storm's cessation. Numerous smaller storms throughout 1997 deposited less precipitation totals on this subwatershed that caused minimal impacts on Q_t. In 1998, Hurricane Bonnie deposited 183 mm of precipitation on the watershed and increased Q_t from 800 m^3 d^{-1} to 51 x 10^3 m^3 d^{-1} within 2 days past this hurricane.

The most severe storm activity occurred in 1999. In January, three fast moving storms (FMS), which passed over the watershed in 1-2 days, collectively deposited 250 mm of precipitation. As will be presented, these three FMS abruptly influenced Q_t and NO_3-N export loads. For the remainder of 1999, 1 TS and 3 successive hurricanes, which passed over the watershed, delivered impressive precipitation amounts that caused extreme hydrological disturbances to the ISW.

Month/year	Storm activity	Precipitation (mm)
August 1997	TS-1	70
August 1998	Hurricane Bonnie	183
January 1999	Fast Moving Storms (FMS)-2, 3, and 4	250
July 1999	TS-2	76
August 1999	Hurricane Dennis	166
September 1999	Hurricane Floyd	392
October 1999	Hurricane Irene	48

Table 1. Dates for tropical storms and hurricanes recorded over the study site and precipitation totals from these storm events.

3.2. Annual Flows from the In-stream Wetland

In accordance with storm activity presented in Table 1, annual Q_t flows varied considerably across three years (Table 2). The lowest annual Q_t occurred in 1997 which corresponds to the lowest annual precipitation total. The annual Q_t, Q_{dr}, and Q_b for 1998 were higher than 1997 which was explained by the higher 1998 precipitation total. The highest Q_t occurred in 1999 in response to 3 FMS, 1 TS and 3 hurricane events (Dennis, Floyd, and Irene) which collectively deposited 326 and 831 mm of precipitation, respectively. In particular, the collective precipitation total from these three hurricanes accounted for 47% of the total annual 1999 precipitation total. Correspondingly, the highest annual Q_{dr} and Q_t flow from the ISW were measured in 1999, which indicates a tremendous amount of runoff entered this ISW in only a 3-month period.

Our flow monitoring results, however, showed that most water exited this ISW during base flow periods. In all three years, Q_b values from the ISW were higher than Q_{dr} with Q_b events accounting for 53 to 75% of the Q_t. The higher Q_b values implies that this ISW receives over 60% of its water supply through ground water recharge. It is not uncommon for stream Q_b to exceed Q_{dr} flows exiting watersheds in the Eastern USA Coastal Plain region [35].

			Q (m^3 x 10^3)		Flow as a percentage of Q_t	
Year	Precipitation (mm)	Q_t	Q_{dr}	Q_b	Q_{dr}	Q_b
1997	1190	1523	554	968	36	64
1998	1410	1926	589	1337	31	69
1999	1756	8980	3226	5754	36	64

Table 2. Annual precipitation near the study area, and total (Q_t), direct runoff (Q_{dr}), and base (Q_b) flows estimated from the in-stream wetland.

3.3. Dissolved N Loads Exported from the In-Stream Wetland

3.3.1. Total Dissolved N Exports

Quantities of N exported from the ISW were expressed as a function of annual Q totals (Table 3) and as annual cumulative totals (Figure 3). The annual total quantity of both NO_3-N and NH_4-N exported from the ISW was variable. For instance, annual NO_3-N export ranged from 4910 to 24,255 kg and NH_3-N loads ranged from 778 to 1852 kg (Table 3). In 1997, more NO_3-N and NH_4-N were exported during Q_b which is consistent with the majority (64%) of ISW outflow occurring as Q_b (Table 2). The monthly amounts of NO_3-N exported increased gradually during 1997 (Figure 3). There was a pulse in NH_4-N exported during June to August 1997 in response to increased outflows from TS-1 (Table 1). In 1998, more NO_3-N was exported from the ISW than in 1997, which is reflective of the higher annual precipitation total (Table 2) and corresponding Q_t (1926 m^3 x 10^3). The cumulative monthly export of both NO_3-N and NH_4-N over 1998 was gradual, with no major pulses observed (Figure 3). It should be noted that there were small pulses of NO_3-N and NH_4-N exported from the ISW (144 and 37 kg, respectively) over the three days that Hurricane Bonnie delivered 183 mm of precipitation (Table 1). But, these pulses were indistinguishable in the cumulative export curves (Figure 3).

		Mass exported (kg)		Mass as % of total	
Year	**Q_t**	**Q_{dr}**	**Q_b**	**Q_{dr}**	**Q_b**
			NO_3-N		
1997	4910	1435	3475	29	71
1998	9961	2740	7176	28	72
1999	24255	7192	17063	30	70
			mean	29	71
			NH_4-N		
1997	778	333	445	43	57
1998	453	157	296	35	65
1999	1852	694	1158	37	63
			mean	38	62

Table 3. Annual mass estimates of dissolved NO_3-N and NH_4-N exported from the in-stream wetland during total (Q_t), direct runoff (Q_{dr}), and base (Q_b) flow events.

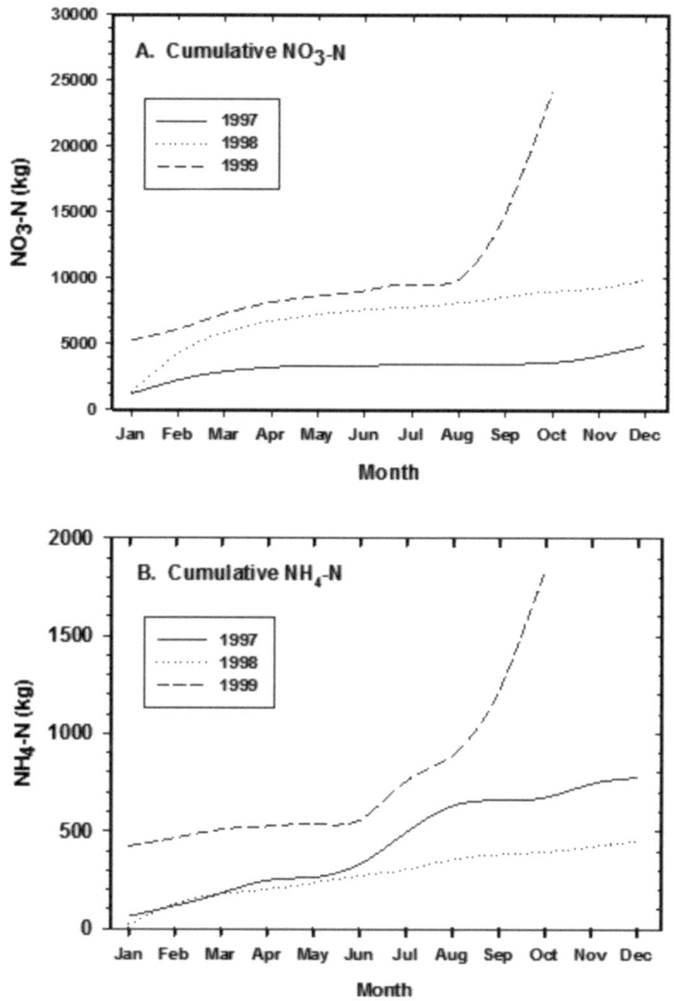

Figure 3. Cumulative annual dissolved NO_3-N (A) and NH_4-N (B) mass exported from the ISW.

3.3.2. *Base Flow Dissolved N Export*

In spite of the storm activity, most of the NO_3-N and NH_4-N as a percentage of the Q_t were exported from the ISW during Q_b events (Table 3). Annual base flows accounted for 70 to 72% and 57 to 65%, respectively, of NO_3-N and NH_4-N exported. Others have reported similar results that more N is exported during Q_b than Q_{dr} events [35, 36].

The greater N masses transported during Q_b events implies that most of the N entering this ISW occurred during ground water recharge. This deduction is supported in reference [37] which reported that NO_3-N discharge rates were greater for watersheds having soil or aquifer characteristics favoring NO_3-N infiltration and subsurface flow. The sandy, agricultural soils in the southeastern Coastal Plain region indeed have these soil and geological features. Stone et al. [38] reported that in localized areas of the HMR watershed, farms that over-applied animal manure effluent had ground water NO_3-N concentrations exceeding 20 mg L^{-1}. This supports the contention that rapid infiltration coupled with lateral flow can facilitate NO_3-N leaching out of the root zone that later emerges into streams. Therefore, more NO_3-N transport during stream Q_b events implies that NO_3-N infiltration is a more predominant mechanism of N transport than direct runoff. The NO_3-N transport mechanisms between soil and ground water, however, can be dramatically altered by storm events that favor NO_3-N movement into the ISW via flooding or as direct runoff.

3.3.3. Direct Runoff Dissolved N Export

Over three years, NO_3-N and NH_4-N exported during Q_{dr} events from this ISW accounted for 29 to 43% of the total annual mass loads, respectively (Table 3). In 1999, however, there was a large increase in both N forms exported during Q_{dr} that prompted a closer inspection of results. The results gathered in 1999 were scrutinized to report daily precipitation totals along with a hydrograph sorted by flow events and coupled with daily NO_3-N export estimates (Figure 4). Nitrate was selected for presentation instead of NH_4-N simply due to the sheer mass export size differences. During January 1999, three FMS events collectively delivered 250 mm of precipitation within the subwatershed. These storms caused a moderate rise is Q_t, but also a huge increase in NO_3-N was exported (5248 kg). The large spike in NO_3-N load exported during January was attributed to abundant water column NO_3-N concentrations (monthly mean = 6.17 ± 1.55 mg L^{-1}) exiting the ISW. For comparison, monthly mean NO_3-N concentrations during the warmer months of June to September were much lower at 0.84 to 3.13 mg L^{-1}, respectively. The higher NO_3-N concentrations during the winter month of January are consistent with reduced N assimilation from slowed plant uptake and denitrification processes in a Coastal Plain wetland [23].

Precipitation amounts during February through July were not eventful as noted by the small spikes in the Q hydrograph and NO_3-N chemograph (Figures 4 B & C). Toward the end of the traditional hurricane season, however, large amounts of precipitation (Figure 4 A, 831 mm) from the three successive hurricanes fell on the watershed causing gross hydrologic disturbances and corresponding higher Q_t (Figure 4 B). These hurricanes also induced more Q_{dr} as noted by the numerous spikes from September to October (Figure 4 B). In turn, the NO_3-N chemograph responded to these hydrologic disturbances by showing dramatic increases in mass export. In fact, between October 1st and 24th, which had the subwatershed draining precipitation from Hurricanes Floyd and Irene, almost 9200 kg of NO_3-N was exported from the ISW. Our results from 1999 imply that N storage by this ISW can be overwhelmed through hydrologic disturbances from storms during both non-hurricane and hurricane seasons.

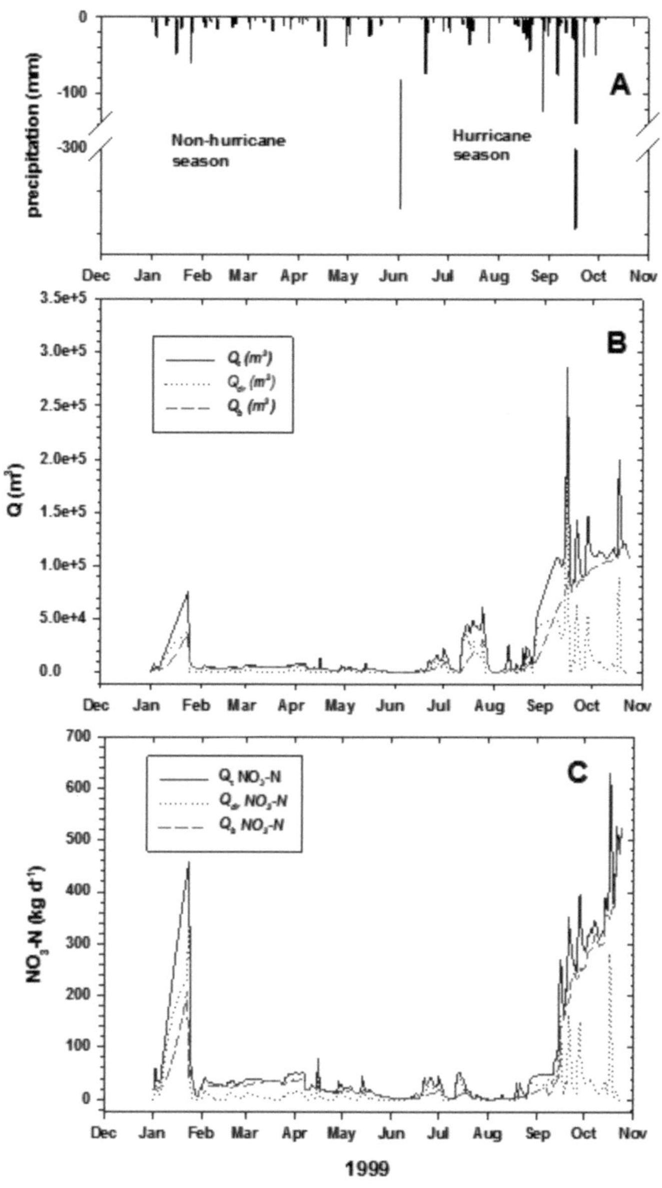

Figure 4. Daily precipitation, outflows and dissolved NO_3-N exported from the ISW during 1999.

This ISW was found to release dissolved N due to storm precipitation and subsequent outflows. As more outflow occurred, correspondingly higher NO_3-N and NH_4 masses were transported out. To quantify the strength of this assertion, linear regression analyses were used to determine if daily NO_3-N and NH_4-N mass exported from the ISW were dependent on daily Q_t (Table 4). Regression analyses revealed that the exported NO_3-N and NH_4-N mass loads were significantly dependent on Q_t; however, the strength of the predicted changes was weak to strong (r^2 between 0.21 and 0.92). There was a predictably stronger ability in 1998 to estimate daily N loads as a function of Q_t, but the strength of the relationship weakened in 1997 and 1999. Arguably, there are other processes not considered in this relationship that will influence N export loads from this ISW such as sediment associated N losses, N sequestered by biological systems, or the rate of N exchange between the water column and sediments. Nonetheless, this study demonstrates that this ISW's ability to store N forms can be compromised as outflow increases.

	year	n	Slope (x 10^{-8})	Y_{int}	r^2	P
NO_3-N						
	1997	365	278	1.84	0.40	<0.001
	1998	364	510	0.21	0.92	<0.001
	1999	297	224	13.92	0.62	<0.001
NH_4-N						
	1997	365	67.5	-0.68	0.21	<0.001
	1998	365	25.9	-0.13	0.71	<0.001
	1999	294	16.5	1.27	0.35	<0.001

Table 4. Linear regression response between total (Q_t as L d^{-1}) outflow and pooled daily dissolved nitrate (NO_3-N) and ammonium (NH_4-N) mass loads exported (kg d^{-1}) from the in-stream wetland.

3.4. Nitrogen Storage Pools Within the In-stream Wetland

Nitrogen storage processes within a wetland involve many pathways distributed between several biotic and abiotic sources [20]. In this study, abiotic N storage pools were examined such as N associated with sediments (as TKN) and soluble N forms in the water column and sediment pore water. The relative changes in N concentrations associated with sediments and between pore water and the water column provided a snapshot of N accretion and losses. The NO_3-N and NH_4-N concentrations were measured in the water column and sediment pore water using passive samplers (peepers) at three locations within the ISW, in sediments below the beaver dam (Figure 1), and by collecting sediments near these same locations and determining their TKN concentrations (Table 5).

At both inlet locations, there were < 11 mg L^{-1} of NO_3-N in the water column and, by 1999, the concentrations declined to < 7 mg L^{-1} (Figure 5). The general decline in water column NO_3-N at both inlets was likely due to flushing of the stream sediments and riparian areas due to storm events during 1997, 1998 and 1999 [26]. The relative N concentrations between the water column and sediment pore water are important parameters to compare because they can be used to determine the gradient direction and subsequent nutrient movement response across the interface [39-41]. In all situations, sediment pore water NO_3-N concentrations at these inlets are lower than the water column, implying that the concentration gradient would cause NO_3-N movement across the sediment water column interface resulting in transfer into the sediments. The NO_3-N equilibrium exchange into the sediment phase is probably accentuated through denitrification by microbial communities [19, 23]. As more NO_3-N is consumed by denitrifying communities in the sediments, a concentration gradient will be created and cause its movement downward. This is a plausible explanation since the NO_3-N concentrations declined rapidly as a function of sediment depth. In fact, the NO_3-N concentrations declined to near zero between 1 and 3 cm deep at the East inlet (Figure 5).

At the Mid-point of the ISW, the water column NO_3-N concentrations over three years showed similar decline behavior as measured at the East inlet. In that, higher NO_3-N concentrations occurred in 1998 than in 1997. But both of these sites (East inlet and Mid-point) experienced a sizable NO_3-N concentration decline in 1999. Similarly, NO_3-N concentrations declined substantially within a few cm below the water column-sediment interface. Likewise to the NO_3-N concentration gradient that occurred at both inlets, the lower sediment pore concentrations at the Mid-point over the three years would cause an equilibrium gradient shift resulting in its transfer into the sediment pore water. The downward NO_3-N movement into the sediments would again cause it to be consumed by denitrifying organisms and hence its concentration declines in pore water (Figure 5).

At the outlet, water column NO_3-N concentration declined to < 0.25 mg L^{-1} during 1997 and 1999, but in 1998 was as high as 3 mg L^{-1}. Sediment pore water at the outlet during the three years was almost devoid (< 0.5 mg L^{-1}) of NO_3-N, suggesting that intensive denitrification prompted its removal from the water column.

Monitoring NO_3-N concentrations in the water column and sediment pore water during the 2-week window along a flow gradient illustrated that this ISW had episodes showing outflowing NO_3-N concentration reductions. Nitrate consumption by denitrifying organisms in the ISW sediments is suspected as being involved with its removal. Low outflow Q_t during the summer months could have facilitated NO_3-N diffusion across the sediment interface and subsequent N removal by denitrification. This removal mechanism was probably overwhelmed by storm events that accelerated NO_3-N export with outflowing water faster than could diffuse across the sediment interface and be consumed by the denitrifying microbial community.

Monitoring NH_4-N concentrations using peepers along the ISW flow continuum indicated some different dynamics relative to NO_3-N. The NH_4-N concentrations in the water column ranged from < 1 to as high as 6 mg L^{-1} (Figure 6). At all four locations, the NH_4-N concentrations declined in a direction towards the sediment-water column interface. Szögi et al. [19] reported in flooded sediment, both nitrification and denitrification can occur at the same time. Similarly, the noted decline may be due to NH_4-N being nitrifying by microbial communities and suggests the presence of aerobic zones above the interface allowing for organisms to conduct this oxidation [42]. The NO_3-N product from this reaction could later diffuse down into the anaerobic zones in the sediments and be removed through denitrification reactions as shown by the NO_3-N depletion with sediment depth (Figure 5).

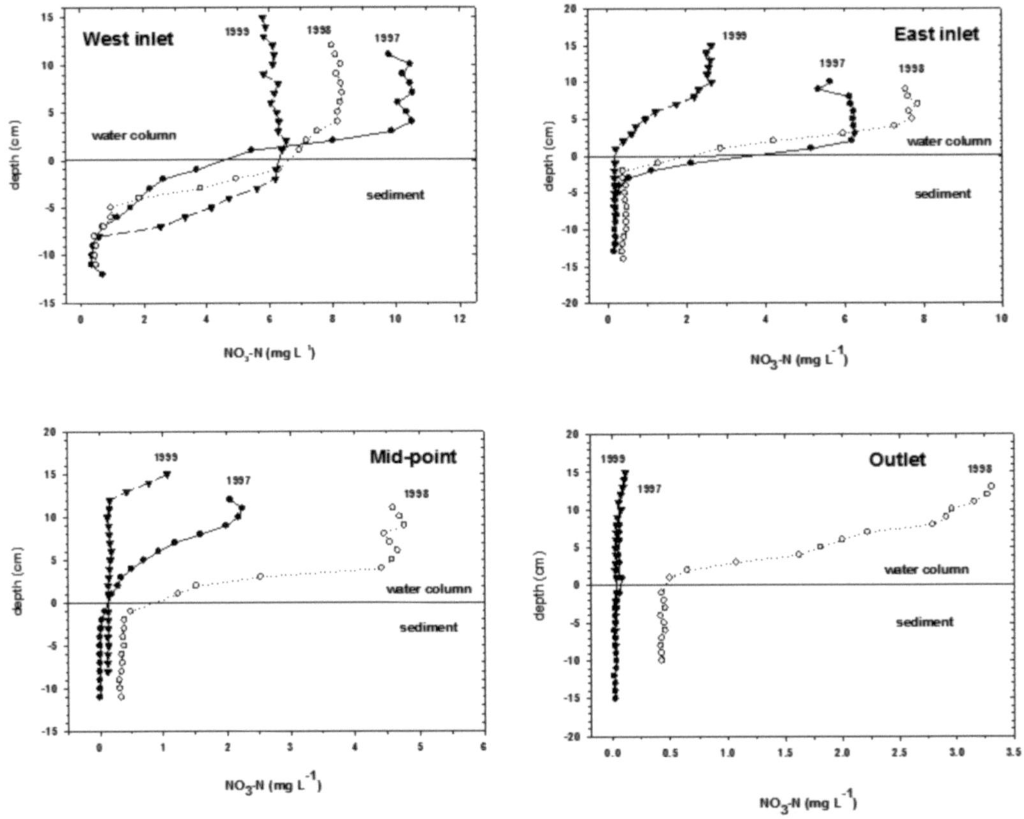

Figure 5. Nitrate (NO_3-N) concentrations measured in the water column and sediment at location in the in-stream wetland (1997 to 1999).

Once oxygen diffusion into the sediments becomes limited, however, the sediments will become more anaerobic thus favoring NH_4-N accumulation. Ammonium does accumulate in the sediment pore water at all four locations (Figure 6). Their concentrations ranged from < 1 to almost 11 mg L^{-1}. Accumulated NH_4-N in sediment pore water can move upward across the interface into the water column [19, 43]. Upward diffusion of NH_4-N would occur in this system because the water columns have lower concentrations than the sediment pore water. Predictability, sediments in this ISW can act as an N source for the overlying water column because of NH_4-N storage in sediments. Once diffused upward into aerobic zones, it can be nitrified to NO_3-N during low flow conditions. During higher flow conditions, NO_3-N would transfer with outflowing water and move down system. The balance between NH_4-N upward movement, NO_3-N consumption, or removal will be a function of stream flow as well as other processes (plant uptake, ammonia volatilization, etc.) known to affect N dynamics in wetlands [19, 20].

Location	1997	1998	1999
	------------	TKN (mg kg^{-1})	------------
East inlet	665	617	2536
West inlet	237	203	137
Midpoint	3371	4337	548
Outlet	854	291	322

Table 5. Sediment Total Kjedhal Nitrogen (TKN) concentrations measured at various in-stream wetland locations.

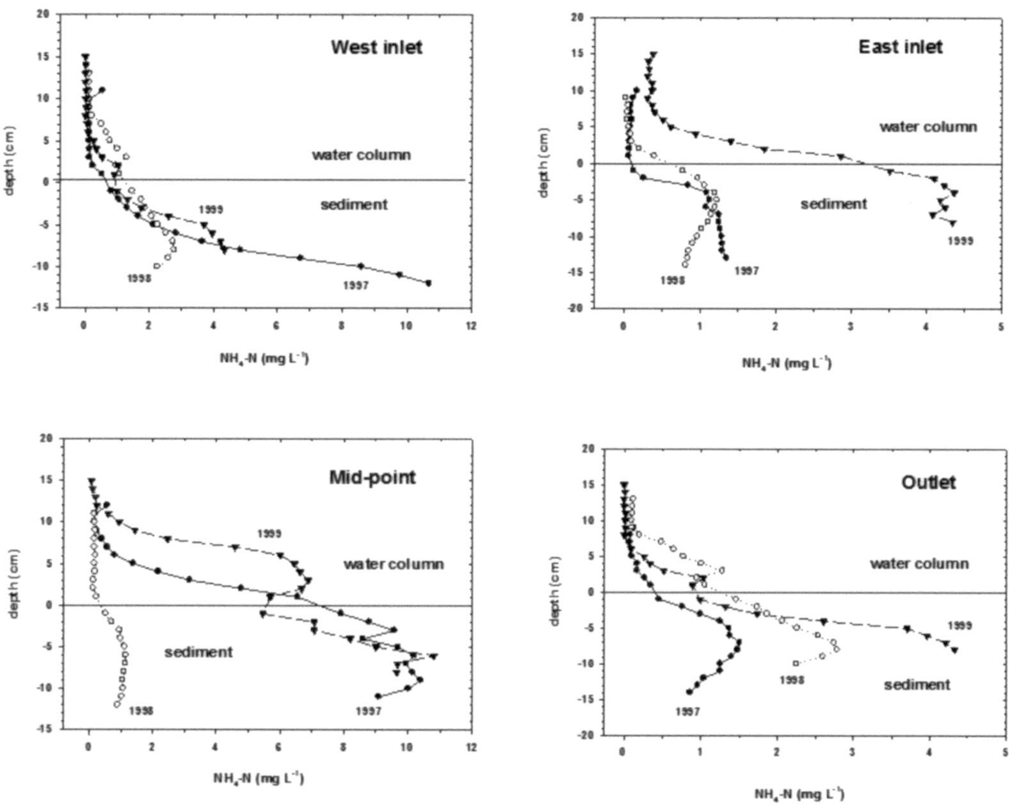

Figure 6. Ammonium (NH_4-N) concentrations measured in the water column and sediment at locations in the in-stream wetland (1997 to 1999).

The TKN content of the ISW sediments was measured in core samples collected when peepers were installed (Table 5). These samples provided a snap-shot of TKN contents in sediments along the flow continuum. The East inlet had higher TKN contents than the West inlet during all three sampling years. The increase in TKN content in sediments at the West inlet between 1997 and 1999 was sizable (almost 4-fold). This may be attributable to relocation of sediment enriched N from upstream riparian sources and/or N leakage from a nearby retired swine lagoon as a result of storm activity 1 to 2 months before collection [28]. On the other hand, the ISW can store N associated with sediments because some of the highest TKN contents were measured in 1997 and 1998 at its Mid-point. Between 1998 and 1999 at the Mid-point, however, there was almost an 8-fold decline in TKN content. Likewise, there was a 3-fold decline in TKN concentration at the outlet. The noted shift in sediment TKN contents implies that water flowing through the East inlet was an apparent N source for this ISW and that storm activity between 1998 and 1999 considerably reduced sediment TKN contents. As outlined [26], this ISW was inundated in 1999 with a tropical storm and Hurricane Dennis a few months before sediment collection and peeper installation. It is plausible that shifts in sediment TKN concentrations were in response to hydrologic disturbances created by these storms. As shown in Fig. 4 C, Hurricane Dennis in mid-Sep. 1999 greatly accelerated NO_3-N export due to the increase in outflow (Figure 4 B).

4. Conclusions

Wetlands are important landscape features in the Southeastern USA Coastal Plain region with respect to water quality because they can act as N sinks. Their role as a natural water filter can be reversed, however, because they can act as an N source when precipitation events from tropical storms and hurricanes increase runoff and flooding causing more N transfer into the wetland. If the wetland's ability to retain this additional N is compromised, then N is released into downstream aquatic ecosystems. This study examined a North Carolina ISW's ability to retain as well as release N during different stream outflow events over a 3-yr period and also examined shifts in N storage pools using passive samplers and sediment TKN concentrations. Over 3 yr, most NO_3-N and NH_4-N mass loads were exported during Q_b events implying that under non-storm conditions the ISW was capable of retaining N. Results obtained in 1999, in contrast, revealed that three successive hurricane events grossly accelerated outflow Q and N releases. Not only was N export accelerated during the typical hurricane season, but a few FMS events during non-hurricane seasons also caused a large pulse of NO_3-N export. During the non-hurricane seasons, this study showed that NO_3-N mass loads are augmented when stream NO_3-N concentrations are higher and by reduced N uptake via biological communities.

Regression analyses showed that annual NO_3-N and NH_4-N mass releases from this ISW were significantly correlated with Q_t flows, but the prediction of N export was weak to strong. This implies a modest ability to predict NO_3-N and NH_4-N export with Q_t.

As N flowed into this ISW, internal and external forces caused by chemical equilibria, flow dynamics, and biological uptake can promote shifts in the abiotic and biotic N storage pools. This study examined only abiotic N storage pools such as bound to sediments or shifted between the water column and pore water N pools. These N storage pools were subject to perturbation by external and internal dynamics forces because both sediment TKN, pore water NO_3-N and NH_4-N concentrations shifted in response to hydrologic disturbances from storm events. Long-term monitoring of the ISW revealed that this ISW's ability to store/ release N was dependent on outflow Q characteristics, sediment movement, and N exchanges between sediments and pore water. This ISW can store N; however, most storage occurred during low out flow(Q_b) situations because flushing of stored N in pore water and associated sediments was minimal. When outflow is accelerated, hydrologic disturbances can flush N out because of equilibria shifts between pools as well as dispersion of N-enriched sediments. It is during these large hydrologic fluctuations associated with storm events when this ISW acts as a pronounced N source for downstream aquatic ecosystems.

Acknowledgements

Sincere gratitude is expressed to Dr. Maurice Cook, Dr. Frank Huminek (deceased), Mr. Mark Rice, and the North Carolina Cooperative Extension Service for project coordination and monitoring. This research project was partially funded by CSREES Grant No. 58-66570-11 entitled "Management practices to reduce non-point source pollution on a watershed basis". [†]Mention of a specific product or vendor does not constitute a guarantee or warranty of the product by the US department of Agriculture or imply its approval to the exclusion of other products that may be suitable.

Author details

J. M. Novak[1*], A. A. Szogi[1], K.C. Stone[1], X. Chu[2], D. W. Watts[1] and M. H. Johnson[1]

*Address all correspondence to: jeff.novak@ars.usda.gov

1 USDA-ARS-Coastal Plains Soil, Water, and Plant Research Center, Florence, South Carolina, U. S. A.

2 Department of Civil Engineering, North Dakota State University, North Dakota, U. S. A.

References

[1] Paerl, H. W., Rudek, J., & Mallin, M. A. (1990). Stimulation of phytoplankton production in coastal waters by natural rainfall input: Nutritional and trophic implications. *Mar. biol.*, 107, 247-254.

[2] Rudek, J., Paerl, H. W., Mallin, M. A., & Bates, P. W. (1991). Seasonal and hydrological control of phytoplankton nutrient limitation in the lower Neuse River Estuary, North Carolina. *Mar. ecol. prog. ser.*, 75, 133-142.

[3] Carpenter, S. R., Caraco, N. F., Correll, D. L., Howarth, R. W., Sharpley, A. N., & Smith, V. H. (1998). Nonpoint pollution of surface waters with phosphorus and nitrogen. *Ecol. appl.*, 8, 559-568.

[4] Howarth, R. W., Sharpley, A. N., & Walker, D. (2002). Sources of nutrient pollution to coastal waters in the United States: Implications for achieving coastal water quality. *Estuaries*, 25, 656-676.

[5] Burkholder, J. M. (1998). Implications of harmful microalgae and heterotrophic dinoflagellates in management of sustainable marine fisheries. *Ecol. appl.*, 8, 537-562.

[6] Bricker, S. B., & Ferreira, J. (2008). ASSESTS eutrophication assessment: Method and application. Available: http://www.loicz.org/imperia/md/content/loicz/science/assests_euthropphication_assessment.pdf. Accessed 2012 May 3.

[7] Van Dolah, P. R., & Anderson, G. (1991). Effects of Hurricane Hugo on salinity and dissolved oxygen condition in the Charleston harbor. *Estuary. j. coastal res.*, 8, 83-94.

[8] Winn, R., & Knott, D. (1992). An evaluation of the survival of experimental populations exposed to hypoxia in the Savannah River Estuary. *Mar. ecol. prog. ser.*, 88, 161-179.

[9] Paerl, H. W., Pickney, J. L., Fear, J. M., & Peierls, B. L. (1998). Ecosystem responses to internal and watershed organic matter loading: Consequences for hypoxia in the eutrophying Neuse River Estuary, North Carolina, USA. *Mar. ecol. prog. ser.*, 166, 17-25.

[10] Kellogg, R. L., Lander, C. H., Moffit, D. C., & Gollehon, N. (2000). Manure nutrients relative to the capacity of croplands and pasturelands to assimilate nutrients: Spatial and temporal trends for the United States. USDA-NRCS-ERS Public. NPS00-05790, United States Dept. of Agriculture, Washington, D.C.

[11] Boyer, E. W., Goodale, C. L., Jaworski, N. A., & Howarth, R. W. (2002). Anthropogenic nitrogen sources and relationships to riverine nitrogen export in the northeastern USA. *Biogeochemistry*, 57, 137-169.

[12] Mallin, M. A., & Cahoon, L. B. (2003). Industrial animal production-A major source of nutrient and microbial pollution to aquatic ecosystems. *Population and environ.*, 24, 369-385.

[13] Mallin, M. A., Posey, M. H., Mc Iver, M. R., Parsons, D. C., Ensign, S. H., & Alphin, T. D. (2002). Impacts and recovery from multiple hurricanes in a Piedmont Coastal Plain river system. *Bioscience*, 52, 999-1010.

[14] NCDA (2009). Statistics-Livestock, Poultry and Dairy. North Carolina Department of Agriculture, Agricultural Statistics Division. Raleigh, North Carolina. Available: http://www.ncagr.com/stats/index.htm.Accessed 2012 May 3.

[15] Mallin, M. A., Cahoon, L. B., Parsons, D. C., & Ensign, S. H. (2001). Effect of nitrogen and phosphorus loading on plankton in Coastal Plain blackwater streams. *J. freshwater ecol.*, 16, 455-466.

[16] Kellogg, R. L. (2000). Potential priority watersheds for protection of water quality from contamination by manure nutrients. Proc. Animal Residuals Manage. 12- 14 Nov 2000 Conf Water Environ. Feder. Alexandria, Virginia

[17] Galloway, J. N., Dentener, F. J., Capone, D. G., Boyer, E. W., Howarth, R. W., Seitzinger, S. P., Asner, G. P., Cleveland, C. C., Green, P. A., Holland, E. A., Karl, D. M., Michaels, A. F., Porter, J. H., Townsend, A. R., & Vöosmarty, C. J. (2004). Nitrogen cycles: Past, present, and future. *Biogeochemistry*, 70, 153-226.

[18] Peterson, B. J., Wollheim, W. M., Mulholland, P. J., Webster, J. R., Meyer, J. L., Tank, J. L., Martí, E., Bowden, W. B., Valett, H. M., Hershey, A. E., Mc Dowell, W. L., Dodds, W. K., Hamilton, S. K., Gregory, S., & Morrall, D. D. (2001). Control of nitrogen export from watersheds by headwater streams. *Science*, 292, 86-90.

[19] Szögi, A. A., Hunt, P. G., & Humenik, F. J. (2003). Nitrogen distribution in soils of constructed wetlands treating lagoon wastewater. *Soil sci. soc. am. j.*, 67, 1943-1951.

[20] Kadlec, R. H., & Knight, R. L. (1996). Treatment wetlands. Lewis Publ., Boca Raton, Florida p., 365.

[21] Stober, J. T., O'Connor, J. T., & Brazos, B. J. (1997). Winter and spring evaluations of a wetland for tertiary wastewater treatment. *Water environ. res.*, 69, 961-968.

[22] Groffman, P. M., Dorsey, A. M., & Mayer, P. M. (2005). N processing within geomorphic structures in urban streams. *J. n. am. benthol. soc.*, 24, 613-625.

[23] Hunt, P. G., Stone, K. C., Humenik, F. J., Matheny, T. A., & Johnson, M. H. (1999). In-stream wetland mitigation of nitrogen contamination in a USA Coastal Plain stream. *J. environ. qual.*, 28, 249-256.

[24] Stone, K. C., Hunt, P. G., Novak, J. M., Johnson, M. H., Watts, D. W., & Humenik, F. J. (2004). Stream nitrogen changes in an eastern Coastal Plain watershed. *J. soil water. conserv.*, 59, 66-72.

[25] Bales, J. D. (2003). Effects of Hurricane Floyd inland flooding, September-October 1999, on tributaries to Pamlico Sound, North Carolina. Estuaries , 26, 1319-1328.

[26] Novak, J. M., Szögi, A. A., Stone, K. C., Watts, D. W., & Johnson, M. H. (2007). Dissolved phosphorus export from an animal waste impacted In-stream wetland: Response to tropical storm and hurricane disturbance. *J. environ. qual.*, 36, 790-800.

[27] Daniels, R. B., Buol, S. W., Kleiss, H. J., & Ditzler, C. A. (1999). Soil systems of North Carolina. *Res. Bull.*, 314. North Carolina, State Univ., Raleigh, NC., p118.

[28] Novak, J. M., Stone, K. C., Watts, D. W., & Johnson, M. H. (2003). Dissolved phosphorus transported during storm and base flow conditions from an agriculturally intensive southeastern Coastal Plain watershed. *Trans. ASAE*, 46, 1355-1363.

[29] State Climate Office of North Carolina. (2009). Daily rainfall recorded at Warsaw, North Carolina. State Climate office of North Carolina, Raleigh, North Carolina.

[30] Lyne, V. D., & Hollick, M. (1979). Stochastic time-variable rainfall-runoff modeling. Hydrology and Water Resource Symp. 10-12 Sept. 1979. Institute of Engineers Australia. Perth, Australia.

[31] Nahatan, R. J., & Mc Mahon, T. A. (1990). Evaluation of automated techniques for baseflow and recession analysis. *Water resour. res.*, 26, 1465-1473.

[32] U.S. Environmental Protection Agency (1983). Methods for chemical analysis of water and wastes. EPA-600/4-79-020. Environmental Monitoring and Support Lab, Office of Research and Development, USEPA, Cincinnati, Ohio.

[33] Novak, J. M., Stone, K. C., Szögi, A. A., Watts, D. W., & Johnson, M. H. (2004). Dissolved phosphorus retention and release from a Coastal Plain in-stream wetland. *J. environ. qual.*, 33, 394-401.

[34] National Oceanic and Atmospheric Association (2012). Hurricane Seasonal Patterns. Available: http://www.nhc.noaa.gov.Accessed 2012 May 4.

[35] Jordan, T. E., Correll, D. L., & Weller, D. E. (1997). Relating nutrient discharges from watersheds to land use and stream flow variability. *Water resour. res.*, 33, 2579-2590.

[36] Owens, L. B., Edwards, W. M., & Keuren, R. W. (1991). Baseflow and stormflow transport of nutrients from mixed agricultural watersheds. *J. environ. qual.*, 20, 407-414.

[37] Brenner, F. J., & Mondok, J. J. (1995). Nonpoint source pollution potential in an agricultural watershed in northwestern Pennsylvania. *Water resour. bull.*, 31, 1101-1112.

[38] Stone, K. C., Hunt, P. G., Johnson, M. H., & Matheny, T. A. (1998). Nitrate-N distribution and trends in shallow groundwater on an Eastern Coastal Plains watershed. *Trans. ASAE*, 41, 59-64.

[39] van Kessel, J. F. (1977). Removal of nitrate from effluent following discharge on surface water. *Water res.*, 11, 533-557.

[40] Reddy, K. R., Kadlec, R. H., Flaig, E., & Gale, P. M. (1999). Phosphorus retention in streams and wetlands: A review. *Crit. rev. environ. sci. tech.*, 29, 83-146.

[41] Birgand, F., Skaggs, R. W., Chescheir, G. M., & Gilliam, J. W. (2007). Nitrogen removal in streams of agricultural catchments-A literature review. *Crit. rev. environ. sci. tech.*, 37, 381-487.

[42] Reddy, K. R., & Patrick, W. H., Jr. (1984). Nitrogen transformations and loss in flooded soils and sediments. *Crit. rev. environ. control*, 13, 273-309.

[43] Reddy, K. R., Patrick, W. H., & Jr Phillips, R. E. (1976). Ammonium diffusion as a factor in nitrogen loss from flooded soils. *Soil sci. soc. am. proc.*, 40, 528-533.

Chapter 8

Meeting the Medical and Mental Health Needs of Children After a Major Hurricane

Robert C. Gensure and Adharsh Ponnapakkam

Additional information is available at the end of the chapter

http://dx.doi.org/10.5772/51750

1. Introduction

The large number of hurricanes in the Gulf Coast region of the United States after the year 2000 has been a great stimulus for research in health effects and health care delivery after a major hurricane. Much of this study has been devoted to children, with over 200 publications on the subject, nearly three quarters of which were published after Hurricane Katrina in New Orleans, LA (2005). Importantly, this recent research has focused not only on meeting medical health care needs of children, but also on meeting mental health needs which might be unique to the aftermath of a major disaster. This chapter will provide a comprehensive review of this literature, providing a resource for health care providers who are coordinating and providing care after a major hurricane.

The major categories for review are of course medical health and mental health. Within medical health, issues of mortality, acute injuries, ongoing care of existing medical conditions, epidemics, toxic exposures, and effects on long-term health care will be discussed. Regarding mental health, acute and chronic post-traumatic stress disorder and serious emotional disturbance will be reviewed, as well as the impact on other children who might not meet full criteria for these disorders. Impact on children with ongoing mental health needs will also be reviewed. Importantly, success rates and overall impact of therapeutic interventions will be discussed, as these are critical for future hurricane events.

2. Medical health

Issues of medical health needs after a major hurricane can be divided in those in the immediate post-disaster response and ongoing needs which may continue for several years. While acute health care needs and delivery systems in the form of disaster response teams have received considerable attention, those related to more chronic needs are only beginning to be recognized as issues, as needs have arisen after the most recent events. Individuals with chronic medical conditions face special challenges, both acutely and long-term. Toxic exposures related to hurricanes are considered separately. Preparedness is considered last, as improvements in tracking of health care needs which follow a disaster have been leading to revisions in the planning process at all levels.

2.1. Mortality

Mortality statistics are among the first health-related complications of hurricanes to be reported. The United States Centers for Disease Control has been tracking and reporting hurricane-related deaths since hurricane Andrew [1]. In fact, early reports regarding health-related outcomes of hurricanes consist almost entirely of mortality statistics [2-5]. Likewise, efforts for health-related disaster planning were directed towards minimizing storm mortality, by early warning and effective evacuation procedures [6]. Reports on hurricane Isabel started to break down hurricane-related deaths by cause, separating those from the storm itself ("Direct"), from those which occurred during the evacuation procedures and in the aftermath of the storm ("indirect") [7]. Importantly, this analysis also revealed that alcohol or drugs was a factor in 28% of storm-related deaths, and that public education efforts should focus on avoiding these during storm events. Unfortunately, there was very little evidence that residents of New Orleans were attentive to these recommendations following Hurricane Katrina – "hurricane parties" remained ubiquitous, and beer was more likely to be taken from convenience stores than water. A summary of hurricane-related deaths in 2004-2005 revealed that 59% of deaths occurred in the postimpact phase, mainly from injury, electrocution, and carbon monoxide poisoning [8]. After hurricane Ike in Texas, carbon monoxide deaths (13) exceeded those from drowning (8) [9]. That said, severe storms can still cause significant injury and mortality, particularly if people are not in reinforced structures [10]. While Hurricane Katrina caused significantly greater overall mortality than these storms, the mortality rates in children were largely unchanged in the pre- and post-storm periods [11]. Of note, the same study notes that infant mortality actually declined, but that this likely was the result of evacuation of high-risk premature infants before and after the storm.

2.2. Injuries

Minor injury is a known risk related to hurricane events, as high winds causes damage to structures, increasing risks for injury of the occupants. As such, injury would be expected to be the overwhelming most common reason for individuals visiting an emergency room or an medical relieve site after a major hurricane. This is recognized in the earliest reports [12], with the reported increased injuries presumed to be sustained during the storm itself. However, a more comprehensive review of hurricane-related health care visits after Hurricane Andrew indicated that the majority of injuries treated occurred after the storm had passed, during the clean-up activities [13]. Even more importantly, the combined number of visits for injuries were only a only a small fraction of those seen for ongoing care of existing medical conditions, as is discussed in the next section.

Disaster medical teams see a particularly high proportion of acute minor injuries in the overall population, as would be expected [14]. Chain saw injuries, presumably related to the clean-up efforts, are common in adults [15]. Interestingly, this same report highlighted a disproportionately large number of insect stings in North Carolina following hurricane Hugo. This has not been reported in series from other hurricane events, and in fact a dermatitis epidemic from moth exposure following hurricane Gilbert was attributed to reduced number of stinging insects (the moth's main predator) from the storm event [16]. In addition to injuries, there were increases seen in other common reasons for visiting the ED for adults, such as cardoivascular disease and asthma [17]. In children, the emergency departments reported increased cases of open wounds, gastroenteritis, and skin infections after Hurricane Andrew [18]. While most of the injuries are minor, a significant proportion are major enough to require hospitalization or surgery [19]. On the other hand, there was no report of increased injuries in children in Jamaica following hurricane Gilbert [20]. Accelerant burn injuries are also seen in greater numbers, likely due to use of gasoline-powered electric generators [21]. Traumatic injuries (falls and cuts) were reported in both residents and relief workers after Hurricane Katrina, although the greater number occurred in residents, presumably because of their much greater numbers [22]. The continued presence of floodwaters after Hurricane Katrina lead to more complications of infections with laceration injuries [23]. Interestingly, in contrast to other flooding events in New Orleans, increased numbers parasitic infections were not reported, presumably because the floodwaters were the result of a salt water rather than fresh water incursion.

Injuries are indeed a prominent feature of hurricane-related health issues which prompt the need to seek medical attention. However, the majority of these injuries are sustained not during the storm itself, but rather during the efforts to clean up and repair damage after the storm. Reports of increased injuries in children are inconsistent, presumably because the children are less likely to be participating in the riskiest clean-up activities.

2.3. Ongoing care for existing medical conditions

A major disaster can cause major disruptions to the health care infrastructure, such that any ongoing health care needs can become difficult to meet. As a result, the majority of health care provided in the acute setting is related to management of ongoing or acute minor illnesses that would normally be cared for by the primary care provider. This has been reported consistently in analyses of several different major hurricane events [13, 24, 25]. Hurricane events with more extensive and widespread damage see a greater proportion of these ongoing healthcare and medication refill visits, reaching as high as 90 percent after Hurricane Andrew and Hurricane Katrina [13, 26]. Interestingly, a similar experience was reported in Korea after Cyclone Nargis [27]. Mobile medical units were established to help meet this need [28].

The types of visits observed for ongoing care were mostly adults who are taking chronic medications for minor ailments (i.e. hypertension) [13, 26]. Other illnesses reported by those seeking care include cardiovascular disease, chronic respiratory disease, ob/gyn conditions, and diabetes [29]. Children are less likely to be on chronic medications for such conditions, and those who are on chronic medications are generally receiving them for less severe conditions, i.e. seasonal allergies. However, those children who do suffer from chronic medical conditions have unique health care challenges in a disaster, as is discussed below.

Following Hurricane Katrina, children with chronic medical problems were found to be at increased risk for adverse events, including worsening of asthma, running out of medications, and experience disruptions in care [30]. Disruptions in care and decreased patient adherence rates were reported in TB clinics as well, for both children and adults [31]. Patients with sickle cell disease who relied on the public hospital system for their care reported greater long-term problems with access to care than did those receiving care primarily through private hospitals [32]. A year after Hurricane Katrina, the overall incidence of chronic medical problems was reported to be increased in children living on the Gulf Coast [33]. This is a surprising finding, given that one might expect families of children with chronic medical conditions to be more likely to relocate to an area where specialized healthcare is more accessible following a major hurricane event. It suggests that children with latent conditions may be more likely to come to clinical attention following a storm event, perhaps as a result of toxic exposures in the post-storm period.

Prior to hurricane Ike, the majority of families with type 1 diabetes mellitus were prepared for at least short-term (3 day) medical needs, although those with lower socioeconomic status were less well prepared [34]. In fact, the families were found to be better prepared for emergency diabetes management than they were for disasters in general. Likewise, metabolic control did not differ significantly overall for patients with type 1 diabetes after Katrina [35]. However, it is worth mentioning from personal experience that my only case of diabetic ketoacidosis complicated by cerebral edema occurred one month after Hurricane Katrina, the severity of presentation was most likely exacerbated by the ongoing disruptions in health care and emergency services. One should also mention that one of the major challenges for children with chronic medical conditions following Hurricane Katrina resulted from the extended evacuation period, such that 3 days worth of medical supplies were quickly exhausted. Special shelters were provided with those with medical needs and their families to help address acute needs of these patients [36], including replenishment of short-term supplies and medications as the evacuation period was extended.

Regarding long-term delivery of healthcare for children with other conditions, rates of patients with cochlear implants actually increased for a major program after Hurricane Katrina [37], while care of patients with craniofacial abnormalities was temporarily disrupted [38]. Some of these improved long-term outcomes may be related to establishment of a system of "medical homes" to coordinate care for children with chronic illnesses [39].

While not necessarily a chronic illness, it is worth noting that newly-diagnosed cancer patients who were currently undergoing treatment were displaced by Hurricane Katrina, according to review of cancer registry data [40]. This represents a special concern, as access to records regarding ongoing chemotherapy regimens can be critical for therapies to be successful and to limit toxicity. Likewise, both Tulane and Louisiana State University had transplant programs for children, and ongoing surveillance and immunosuppressive therapy would be complicated by the transfer of care and inability to access previous records.

Overall, ongoing care for chronic medical conditions becomes a major challenge when the local healthcare system is disrupted by a storm event. Hurricane planners should focus on meeting these needs, ensuring a ready supply of the most commonly used drugs and supplies in treating these conditions, and adequate urgent-care type medical facilities to meet the needs of these individuals.

2.4. Evacuees

While much study has been devoted to administration of care to victims at the site of the disaster, the ability to evacuate large populations en-masse to more distant centers extends the need to provide health care services at these locations as well. Of course, the minor injuries resulting from clean-up activities would not be expected in those who have been evacuated to distant centers. However, issues of displacement represent challenges of their own, as evacuees would have no access to their primary care physicians and limited access to pharmacies. After Hurricane Katrina, people were evacuated to almost 1,400 centers in 27 states [41, 42]. Basic needs reported in shelters included dental care, eyeglasses, acute illness, and ongoing management of chronic illnesses [41-43]. Only a small percentage (3.8%) required hospital referral [42]. Monitoring systems were established in some shelters after Hurricane Katrina to provide health needs for the evacuees and to prevent outbreaks [44]. As evacuees are dispersed from designated centers, impacts can be felt on the local health care system akin to a wave of immigration [45]. Of course, that evacuees are presumably relocated to a region with an intact health care system provides a huge advantage in meeting these individuals' healthcare needs.

2.5. Epidemics

Hurricanes cause damage to infrastructure and can force people into closer living situations (i.e. shelters), increasing the chances of outbreaks of infectious diseases. Highly contagious disorders can become epidemic after hurricanes, as was seen in St. Croix after Hurricane Georges, where 88 cases of conjunctivitis were reported in one public health clinic a week after the hurricane (the previous case rate was 3 per month) [46]. Evacuees from Hurricane Katrina in the Astrodome developed an outbreak of gastroenteritis from Norovirus [47]. Fortunately, improvements in surveillance at that site helped prevent this outbreak from becoming more serious [44].

Waterborn infections or infections with water-dependent vectors can also occur more frequently, particularly when prolonged flooding occurs. Dengue fever outbreaks have been reported in Puerto Rico associated with hurricanes and floods [48]. Haiti also experienced difficulties with mosquito-born diseases (malaria, dengue fever, and West Nile Virus) after Hurricane Jeanne [49]. Importantly, these were the first cases of West Nile Virus reported in Haiti, underscoring the need to consider the possibility of disease outbreaks which are otherwise atypical for the region after a hurricane event. An outbreak of waterborn cholera was seen in India after Cyclone Aila [50]. As an interesting variation on hurricane-related epidemics, dermatitis as a result of contact with the *Hylesia alinda* Druce moth was reported in high numbers in Cozumel, Mexico following hurricane Gilbert [16]. A full survey revealed that 12.1% of the population was affected. The link with the hurricane was death of the moth's natural predators, bees and wasps, which allowed the moth population to overgrow.

Outbreaks of infectious disease are a long-recognized risk following hurricane events, and as such, it is thus not surprising to see that surveillance systems and effective responses are being employed during hurricane relief efforts.

2.6. Toxic exposures

Exposure to toxins following hurricanes can come from many sources. Mould and spores growing in flooded houses can cause respiratory difficulties. Toxins such has heavy metals can be released from damaged buildings. Gas generators running continuously can generate excess carbon monoxide if not adequately ventilated. These unique exposures to toxins need to be considered in planning and response to hurricane events.

2.6.1. Respiratory

Asthma is very common in children, and exacerbation of asthma symptoms by hurricanes and the associated flooding is of great concern. As noted previously, childhood visits to emergency rooms for asthma are increased after Hurricane Katrina [29]. A similar experience was reported in North Carolina after flooding from Hurricane Floyd, which damaged schools [51]. Importantly, increased respiratory complaints were not limited to individuals with a pre-existing diagnosis of asthma, as a survey of the general population revealed increased complaints of both upper and lower respiratory symptoms [52]. A general increase in minor respiratory complaints in children was noted in Jamaica after hurricane Gilbert [20]. Risk factors for respiratory disease identified in the study include roof or window damage, outside mould, dust, and interior flood damage. It is thus not surprising that respiratory complaints are among the most consistently reported in surveys of emergency rooms and shelters, as indicated previously, and responders should be prepared to treat a large number of children presenting with these disorders.

2.6.2. Mould

Contamination of home interiors with mould was of great concern after Hurricane Katrina. Many houses were flooded for a month or more, and interior walls of such homes would typically be entirely coated with mould growth. There was concern if all of the mould could be removed from such homes. While this has generated much discussion in the lay press, only one study actually reported mould as a risk factor for respiratory complaints, and this was linked to outdoor mould [52]. A preliminary study in New Orleans showed surprisingly low respiratory symptoms (based on spirometry) and surprisingly low exposure to mould (based on air sampling), both of which improved quickly over a 2 month period [53]. However, this study was conducted in the Garden District, where there was minimal flooding. Concerns about mould exposure are not limited to hurricanes, of course, and experiences with excess in-home mould exposure after hurricanes have stimulated policy development to provide better control of mould and moisture in all homes [54].

2.6.3. Lead

New Orleans is an old city, and as such there were difficulties with lead contamination and exposure to children prior to Katrina [55]. There was concern that the combined effects of storm damage and subsequent renovation would increase risk of leat exposure further by releasing even more lead into the local environment [56]. Mild increases in soil lead levels after Hurricane Katrina were seen in properties previously treated with fresh soil for heavy lead contamination – the soil did not wash away, but rather was re-contaminated from lead released from damaged buildings or during the renovation process [57]. A broader risk analysis for soil lead levels after Hurricane Katrina yielded similar results [58]. Another random survey showed a high proportion (27%) of New Orleans homes with lead levels in surrounding soil well above recommended levels, and average lead levels were increased from those obtained before Hurricane Katrina [59]. Schoolyards were not spared, raising particular concerns for returning children [60]. Importantly, soil lead levels have been correlated with blood lead levels in New Orleans both before and after the storm [61], indicating that release of lead into the environment can have adverse consequences for children. Increased lead in the soil around heavily damaged homes needs to be considered after a hurricane event, and attention needs to be paid during the rebuilding process that renovations do not worsen the situation.

2.6.4. Carbon monoxide poisoning

Massive, prolonged power failures following a hurricane event will result in increased use of gas-powered electric generators. To avoid theft, these generators are often placed to nearer to homes or within covered porches. A Florida study noted some gas-powered generators were placed in garages, or even inside the homes themselves [62]. Carbon monoxide exhausts can become trapped in the home and create a slow, chronic exposure for the occupants. A CDC investigation in Alabama and Texas found 27 incidents in hurricane-affected areas, resulting in 10 deaths [63, 64]. While the majority of individuals were found to be using generators to power stoves or window-unit air conditioners, one study cited deaths following hurricane Ike related to generator use solely to power television and video games [65]. Of note, carbon monoxide poisoning from other causes, particularly automobile exhaust, occurs year-round, accounting for nearly double the number of cases in Florida from those related to gasoline-powered generators [66]. In response to this important source of post-storm morbidity and mortality, public health measures have been taken to educate people about the risks of use of gasoline-powered generators, and how to take steps to minimize those risks.

2.7. Hospitals

In addition to facing all of these special health issues among the served population, hospitals face challenges of their own in a hurricane setting. Decisions about whether to evacuate patients prior to the storm's arrival, and which patients to evacuate, can be problematic [67]. Power outages can strain back-up generator systems, and evacuations and damage to residences can adversely affect hospital staffing. Major flood events, causing failures of power (primary and backup), running water, and, most importantly, sewerage, forced hospitals in downtown New Orleans to close after Hurricane Katrina. Evacuating remaining patients from these hospitals was especially problematic under the circumstances. A seemingly minor point, but one which is potentially critical for nation-wide public health, is the transfer of bacterial flora between hospitals; patient evacuations raised concerns of spread of multidrug-resistant bacterial strains across the country [68]. Despite these logistical challenges, orderly evacuation of pediatric patients was conducted to other hospitals in the southeast [69, 70]. However, evacuations were not immediate, and the challenges, particularly in ICU settings, of providing continued care in the absence of basic resources, required considerable resourcefulness on the part of hospital staff [71]. Fortunately, hospital staff are trained to work under extreme circumstances, and likely by nature of career choice show remarkable willingness to work in disaster setting. In polls of hospital staff conducted across several states, 78% of respondents indicated they would be willing to work through a hurricane [72]. Overall (personal bias notwithstanding!), hospitals can generally be relied on to perform extremely well during hurricane events, with stockpiles of supplies, emergency power systems, serving as anchor points for health care delivery and for other elements of the hurricane response team.

2.8. Long-term effects on health care delivery

The effects of hurricanes and relief efforts which follow on the local health care infrastructure are quite variable. In response to hurricane Mitch, the American Red Cross undertook a program to improve water and sewerage infrastructure in Central America, resulting in decreased prevalence of diarrheal illnesses from pre-storm rates [73]. Long-term access to health care was also significantly improved in the region as a result of the relief efforts [74]. While federal efforts to promote recovery of the health care system after Hurricane Katrina provided funding and resources to hospitals and community health centers, no assistance was offered to private practice physicians [75]; in fact, temporary funding for mobile health clinics had the opposite effect of creating free competition, further straining the private practice model and leading to long-term shortages in primary care physicians in the region, particularly in the field of obstetrics/gynecology.

Access to vaccination records is of great importance, particularly for hurricane evacuees, as proof of vaccination is generally required for children to be allowed to enter school. In the United States, vaccination records are maintained by the individual states rather than in a common federal repository. Connecting the Louisiana and Houston-Harris County immunization registries allowed recovery and immediate access to childhood immunization records after Hurricane Katrina [76]. This allowed for considerable cost savings, preventing the need for re-vaccination of children because of lost records [77].

2.9. Growth and development

Access to adequate food and water can be limited following a hurricane, and there are concerns regarding growth and development of children under these circumstances. Likewise, the stress of the situation can have direct effects on growth and development as well. Growth in height was significantly reduced in Jamaican children following hurricane Gilbert [20]. Higher levels of malnutrition were reported in several regions of Honduras after hurricane Mitch [78]. Decreased incidence of precocious puberty and increased incidence of pubertal delay was observed in New Orleans after Hurricane Katrina [79]. In all cases, adequate available nutrition was well-documented in the study populations, so it appears likely that other storm-related issues affected the distribution of food in relief efforts, or that other medical factors (i.e. stress) influenced growth and development in children in the post-storm period.

2.10. Maternal-fetal health

Exposure of mothers in the second and third trimester to hurricane conditions resulted in increases in several measures of fetal distress [80]. A significant increase in autism was noted in children with prenatal exposure to hurricanes, particularly in mid-late gestation [81], although it is not clear if the prenatal exposure to the hurricane or the postnatal exposure to the recovery period, with the accompanying increased stress, is responsible. Further effects of disasters on perinatal health, particularly possible increases in spontaneous abortion, have been difficult to document and require further study [82]. Because of these known risks, programs were established by the Organization of Teratology Information Specialists and the National Center on Birth Defects and Developmental Disabilities to assist pregnant and breastfeeding women in relationship to possible exposures, such as infections, chemicals, medications, and stress, after Hurricane Katrina [83].

Newborn screening procedures were disrupted in the first month following Hurricane Katrina, as the state lab was located in New Orleans and badly damaged by the storm. However, concerted efforts to identify any lost screening samples and to locate children with positive results allowed treatment to be initiated in all 10 cases with positive screening results [84]. The newborn screening program was temporarily contracted to a different facility, such that there were no gaps in the newborn screening program.

3. Mental health

Individuals in hurricane-affected areas are often exposed to a wide range of stressors including serious risk of death, property loss, difficulty obtaining basic necessities such as food and clothing, and exposure to violence. In areas where effects linger for many months, individuals are exposed to additional stressors such as forced relocation, difficulty obtaining housing, and prolonged community disruption. Children are particularly at risk to these stressors. The effects of these stressors on children after a major hurricane event are discussed below.

3.1. Posttraumatic Stress Disorder (PTSD)

In the first 3 to 6 months following a hurricane, more than 50% of children exposed to the disaster exhibit symptoms of posttraumatic stress disorder (PTSD), disruptive behaviors, or other manifestations of psychological distress, as seen with Hurricans Hugo, Andrew, and

Katrina [85-87]. PTSD is an anxiety disorder of at least one-month duration that is characterized by symptoms of re-experiencing (e.g., intrusive memories or thoughts), avoidance and/or emotional numbing, and hyperarousal. It is often accompanied by feelings of anxiety and depression, social alienation, and mistrust of family, friends, and systems [88]. Previous research has identified exposure to disaster-related stressors as an important predictor of psychiatric symptoms among youths after natural disasters [88].

In 2008, Marsee found that 63% of students in a sample of 166 students from 9th to 12th grade had symptoms of PTSD 15-18 months post-Katrina, and that high levels of aggression were associated with emotional dysreguation [89]. In 2009, Weems et al. expanded the understanding of post-Katrina symptamology, finding that among 52 children with a mean age of 11 years in the 6–7 months following Katrina, the level of posttraumatic stress symptomatology was related to hurricane exposure, female gender, and level of predisaster anxiety [90]. The last is particularly important, as it suggests that hurricane events can serve as triggers in individuals with latent psychiatric disorders. Other variables, such as separation from a caregiver and evacuating to a shelter were associated with PTSD 2 years post-Katrina in 7-19 year-olds [91]. Caregiver symptamology is also associated with hurricane-related PTSD in children across several age groups [92, 93]. These are staggering numbers, and the extent of the challenge of caring for a major psychological disorder in the majority of the population cannot be overemphasized.

3.2. Factors influencing PTSD

Based on the newly emerging literature, PTSD following Hurricane Katrina was found to be very common, with children's responses to the stress and trauma of the hurricane event associated with other environmental and relational factors as well.

3.2.1. Age

Several studies have suggested that age has an impact on PTSD. For example, McDermott and Palmer (2002) studied 8 to 19 year olds who reported loss associated with a bushfire and reported that depression following trauma is more common in younger children than older children [94]. Similar results were found following Hurricane Hugo [95]. Other studies, however, have found differing results. Most notably, a study of adolescents in 9th-12th grades following Hurricane Katrina showed no relationship between age and traumatic response [89].

3.2.2. Gender

The majority of studies have shown that females are more likely than males to develop PTSD symptoms [95, 96]. Additionally, differences in manifested symptomatology have been described, with girls being more likely to express guilt and other emotional reactions, and boys being more likely to exhibit increased worry, anhedonia, concentration problems, academic problems and other cognitive or behavioral symptoms [95, 97].

http://dx.doi.org/10.5772/51750

3.2.3. Race/ethnicity

Some studies suggest that minority populations are at an increased risk for PTSD [98, 99]. This association has been found following Hurricanes as well. Following Hurricane Hugo, Lonigan et. Al. found that African American youth reported more PTSD symptoms than either Caucasian children or other minority youth [100]. Similar results were found following Hurricane Andrew, with both Hispanic and African American children reporting higher levels of PTSD symptoms than Caucasian children at 7 and 10 months post-hurricane [88]. It is important to note that it is difficult to separate the apparent effects of race/ethnicity from the effects of socioeconomic status, and rural vs urban environment. Because families that are impoverished or live in poor urban environments may be more adversely affected by hurricanes, the roles of socioeconomic status and race in PTSD requires more investigation.

3.2.4. Premorbid anxiety

Some studies have examined the influence of emotional and behavioral problems on PTSD. One prospective study conducted after Hurricane Andrew showed that pre-existing anxiety levels predicted the severity of post-traumatic responses in youth with those showing higher levels of anxiety 15 months predisaster reporting higher levels of post-traumatic symptoms 3 and 7 months after the disaster. Additionally, children exhibiting higher levels of premorbid anxiety were less likely to recover over time [101]. After Katrina, another study found that pre-hurricane trait anxiety was significantly associated with post-hurricane PTSD symptoms [102]. As hurricane events appear to cause stresses and disruptions of normal coping mechanisms, it is not surprising that children with highest pre-storm stress are most likely to develop PTSD.

3.2.5. Timescale

For many children, symptoms after natural disasters are relatively short lived, with substantial decreases occurring during the first year post-disaster [86]. For example, nearly 30% of children exposed to Hurricane Andrew reported severe symptoms of PTSD, defined as 10 or more symptoms, 3 months after the storm. At 7 months posthurricane the prevalence of such symptoms had dropped to 18%, and at 10 months posthurricane, 13% of the children still reported severe symptoms [88].

Identification of factors that distinguish children who experience chronic symptoms from those whose distress is more transient represents an important goal, given its implications for targeting postdisaster interventions. Previous research after other natural disasters has identified female sex, younger age, nonwhite race/ethnicity, parent psychopathology, and degree of stress exposure as predictors of long-term symptom elevation in youths. However, the extremely high prevelance of PTSD in the exposed population, particularly in the immediate post-storm period, also raises the possibility that universal, large-scale interventions, most likely implemented in the school setting, should also be considered.

3.3. Serious Emotional Disturbance (SED)

Serious emotional disturbance (SED) is a term that refers to children and adolescents who have a diagnosable mental disorder that results in significant impairment or decreased role functioning in family, school, or community activities. This disorder often manifests itself through aggression or other behavioral issues. While the prevalence of SED has not traditionally part of a post-hurricane evaluations, the months and years following Hurricane Katrina saw a marked increase in SED prevalence. Indeed, school referrals for mental health evaluation and services made following Hurricane Katrina were overwhelmingly for disruptive behavior disorders [103]. Moreover, the prevalence of SED following Hurricane Katrina was long-lasting: the estimated prevalence of SED among children and adolescents exposed to Hurricane Katrina 18 to 27 months after the storm was around 15%, with nearly 10% of these cases deemed by parents to be directly attributable to the hurricane [104]. And while the prevalence of SED among children and adolescents exposed to Hurricane Katrina declined to 11.5% in the next 12-18 months, this prevalence is still nearly three times greater than the pre-hurricane rate of SED (4.2%) sited in the National Health Interview Survey (NHIS) [105].

Examination of individuals with SED following Katrina indicates that approximately two-thirds of children and adolescents with SED at 18-27 months post-Katrina had recovered by the 12-18 month follow up. This recovery, however, was offset by a high rate of new SED onset during the follow-up period; more than half of SED cases at follow-up were not present during the baseline assessment [106]. Substantially more of these cases were attributable to the Hurricane, suggesting that they represent delayed onset of SED rather than normal trends in SED unrelated to the Hurricane. This delayed onset may be attributable to high levels of ongoing stress, community disruption, and other traumatic events due to slow pace of recovery in the Gulf Coast. The large number late-onset cases of hurricane-related SED indicates that there is a need for surveillance of at-risk populations than was previously thought.

Children and adolescents who experienced deficits in functioning prior to the Hurricane were more likely to have persistent SED than children without prior history. This effect was only present among youth who experienced low to moderate stress during and after the Hurricane; youth exposed to highest levels of stress exhibited elevated rates of SED, and more than one-third of these youth continued to exhibit SED at the follow up time point, regardless of previous decrements in function. Consistent with research on other disorders that imply strong associations between cumulative stressors and mental health problems, these results suggest that the magnitude of stress to which these youths were exposed was sufficient to overwhelm their coping resources. This also highlights the need for mental help for youths who experience high levels of stressors following Hurricanes.

This high prevalence suggests that the long-term, widespread impact of Hurricane Katrina on child mental health occurred on a level otherwise unseen. Further, sociodemographic factors typically associated with psychopathological reactions to natural disasters, such as age and sex [95, 98, 107], are largely unrelated to SED in children exposed to Hurricane Katrina. While the reason for this discrepancy remains unclear, it is hypothesized that the magnitude and timescale of disruption during and after Katrina increased risk for mental health problems in youths across many segments of society. Not surprisingly,

http://dx.doi.org/10.5772/51750

stress exposure is associated strongly with SED, and strength of association varies with traumas experienced. In New Orleans area youth, death of a loved one was most strongly associated with SED. In the remainder of children sampled, physical adversity such as difficulty obtaining food and shelter showed the strongest correlation.

Parent psychopathology has been associated with child psychopathology following major disasters, and was associated with SED following Katrina. This may predispose youth to the development of psychiatric symptoms wither through psychological predispositions or ineffective parenting (12, 29). Social class was negatively associated with SED following Hurricane Katrina. While SED is approximately twice as prevalent in children living in poverty (28), certain characteristics of the Hurricane may explain this. Damage from the Hurricane was greatest in areas of the city with high levels of poverty, and families with fewer resources would have found it more difficult to find stable shelter and other basic needs following the Hurricane. This would lead to increased and prolonged exposure of children living in poverty to a wide variety of stressors not felt by those with more resources.

Identifying the psychological and social mechanisms that underlie this complex set of associations between hurricane-related stressors and SED is an important goal for future research.

Overall, while the overwhelming majority of research into psychological effects of hurricane events has traditionally focused PTSD, research following Hurricane Katrina has shown that SED is an important disorder that needs to be considered as well, particularly in the identified at-risk individuals. That SED can present long after (i.e. more than 2 years) after a hurricane event complicates identification of affected individuals, and underscores the need for long-term surveillance of the at-risk population

3.4. Interventions

A system of psychological first aid has been developed by the National Child Traumatic Stress Network and the Department of Veterans Affairs National Center for Posttraumatic Stress Disorder. This system was employed by providers following hurricanes Gustav and Ike, and was found to improve confidence for providers and improve outcomes in adults and children [108]. For children suffering acutely from PTSD related to injuries, a combination of crisis intervention and family support was found to be most effective. It has also been noted that children possess a greater resilience and recovery power related to such events [109]. Interventions which build coping skills were found to be particularly effective [110], as were more structured cognitive-behavioral therapies [111]. Unfortunately, these therapies are complex and require special training, and given the enormous scope of the problem following a major hurricane event such as Hurricane Katrina, it can be difficult to mobilize enough trained therapist to service the affected individuals.

Elementary school-based interventions by counselors for children showing symptoms of psychological trauma was found to be effective in reducing trauma symptoms in a controlled trial [112]. This approach is particularly of interest as it allows mass application of psychological screening and therapy to a large at-risk population without relying on individuals seeking professional evaluation. Project Fleur-de-Lis was another school-based triage and treatment program showing encouraging results utilizing a system designed to provide uniform access for a large at-risk population [113]. School-based interventions were found to be as effective as therapy through a mental health clinic [114]. Again, even with school-based interventions, the application of these techniques following Hurricane Katrina were limited to pilot studies, as adequate resources were not available to extend these interventions to all schools in the Gulf Coast area.

While effective therapies were identified, there are concerns that mental health programs were not applied globally enough, nor were they sustained in a fashion which would be required by the chronic nature of the disorders affecting children [115]. As noted above, a residual incidence of SED of nearly 15% 2 years after Hurricane Katrina underscores the need for ongoing evaluation and therapy in the at-risk population [116]. The problem is even more severe when one looks at rates of PTSD, which despite previous reports indicating the condition is temporary, remain above 40% 33 months after Hurricane Katrina, indicating that stress from the ongoing recovery efforts may have resulted in prolongation of symptoms [117]. Thus, therapy for PTSD and SED were mostly lacking following Hurricane Katrina because of lack of available mental health resources and lack of funding (estimated at $1,133 per capita expense) for comprehensive therapy for the affected population [118]. There is a clear need to develop modified therapies, particularly for PTSD, which can be implemented by personnel without advanced training in psychotherapy.

4. Conclusion

The occurrence of a large number of major hurricane events around the Gulf Coast region of the United States within a relatively short period of time has led to a dramatic increase in research in health effects of hurricane events. Some surprising results include the need for relief medical staff to be able to treat a large number of minor injuries sustained during recovery efforts, and the need to provide more urgent-care services for individuals with acute illnesses and need for ongoing care for chronic medical problems, both children and adults. As such, response teams should be better prepared to cope with a large number of individuals with more common medical complaints, rather than focusing solely on caring for individuals with major injuries incurred during the storm event. In particular, primary care physicians need to be a major part of disaster response teams, and adequate supplies of commonly-prescribed medications need to be provided as part of the disaster response.

Reports of mental health following major hurricane events reveal a shockingly high incidence of post-traumatic stress disorder and of serious emotional disturbance, affecting the majority of children in most surveys. While effective therapies have been described which can be implemented in a school setting, the capacity of trained staff to apply these therapies is overwhelmed by the scope of the problem. It is neither possible nor practical to expect that trained therapists could be made available to evaluate and treat the entire population of children in an affected area, with treatments continuing beyond 3 years from the storm event. Instead, there is a need to develop therapeutic techniques which can be applied by individuals with less

http://dx.doi.org/10.5772/51750

training, i.e. general physicians, school guidance counselors, or (ideally) by teachers to meet this massive need for psychiatric care following major hurricane events.

Author details

Robert C. Gensure[1,2*] and Adharsh Ponnapakkam[3]

*Address all correspondence to: rgensure@montefiore.org

1 Pediatric Endocrinology, Children's Hospital at Montefiore

2 Albert Einstein College of Medicine

3 Tulane University

References

[1] Centers for Disease Control. (1992). Preliminary report: medical examiner reports of deaths associated with Hurricane Andrew--Florida, August. *MMWR Morb Mortal Wkly Rep.*, 41(35), 641-644.

[2] Centers for Disease Control and Prevention. (1996). Deaths associated with Hurricanes Marilyn and Opal--United States, September-October 1995. *MMWR Morb Mortal Wkly Rep.*, 45(2), 32-38.

[3] Combs, D. L., et al. (1996). Deaths related to Hurricane Andrew in Florida and Louisiana, 1992. *Int J Epidemiol.*, 25(3), 537-544.

[4] Hendrickson, L. A., & Vogt, R. L. (1996). Mortality of Kauai residents in the 12-month period following Hurricane Iniki. *Am J Epidemiol*, 144(2), 188-191.

[5] Lew, E. O., & Wetli, C. V. (1996). Mortality from Hurricane Andrew. *J Forensic Sci*, 41(3), 449-452.

[6] Centers for Disease Control and Prevention. (2004). Preliminary medical examiner reports of mortality associated with Hurricane Charley--Florida. *MMWR Morb Mortal Wkly Rep.*, 53(36), 835-837.

[7] Jani, A. A., et al. (2006). Hurricane Isabel-related mortality--virginia, 2003. *J Public Health Manag Pract*, 12(1), 97-102.

[8] Ragan, P., et al. (2008). Mortality surveillance: 2004 to 2005 Florida hurricane-related deaths. *Am J Forensic Med Pathol*, 29(2), 148-153.

[9] Zane, D. F., et al. (2011). Tracking deaths related to Hurricane Ike, Texas, 2008. *Disaster*, 5(1), 23-28.

[10] Shen, J., et al. (2009). Risk factors for injury during Typhoon Saomei. *Epidemiology*, 892-895.

[11] Kanter, R. K. (2010). Child mortality after Hurricane Katrina. *Disaster*, 4(1), 62-65.

[12] Ranhoff, A. H., Naustdal, H., & Skomsvoll, J. F. (1992). Injuries following a hurricane in Nordmore. *Tidsskr Nor Laegeforen*, 112(30), 3777-3780.

[13] Alson, R., et al. (1993). Analysis of medical treatment at a field hospital following Hurricane Andrew, 1992. *Ann Emerg Med*, 22(11), 1721-1728.

[14] Henderson, A. K., et al. (1994). Disaster medical assistance teams: providing health care to a community struck by Hurricane Iniki. *Ann Emerg Med*, 23(4), 726-730.

[15] Brewer, R. D., Morris, P. D., & Cole, T. B. (1994). Hurricane-related emergency department visits in an inland area: an analysis of the public health impact of Hurricane Hugo in North Carolina. *Ann Emerg Med*, 23(4), 731-736.

[16] Fernandez, G., et al. (1992). Epidemic dermatitis due to contact with a moth in Cozumel, Mexico. *Am J Trop Med Hyg*, 46(5), 560-563.

[17] Hendrickson, L. A., et al. (1997). Morbidity on Kauai before and after Hurricane Iniki. *Prev Med*, 26(5), Pt 1, 711-716.

[18] Quinn, B., Baker, R., & Pratt, J. (1994). Hurricane Andrew and a pediatric emergency department. *Ann Emerg Med*, 23(4), 737-741.

[19] Gagnon, E. B., et al. (2005). In the wake of Hurricane Isabel: a prospective study of postevent trauma and injury control strategies. *Am Surg*, 71(3), 194-197.

[20] Simeon, D. T., et al. (1993). Effects of a hurricane on growth and morbidity in children from low-income families in Kingston, Jamaica. *Trans R Soc Trop Med Hyg*, 87(5), 526-528.

[21] Rainey, S., et al. (2007). The occurrence and seasonal variation of accelerant-related burn injuries in central Florida. *J Burn Care Res*, 28(5), 675-680.

[22] Sullivent, E. E., et al. (2006). Nonfatal injuries following Hurricane Katrina--New Orleans, Louisiana, 2005. *J Safety Res*, 37(2), 213-217, Epub May 12.

[23] Chopra, A. K., et al. (2009). Virulence factor-activity relationships (VFAR) with specific emphasis on Aeromonas species (spp.). *J Water Health*, 7(1), S29-S54.

[24] Leonard, R. B., Spangler, H. M., & Stringer, L. W. (1997). Medical outreach after hurricane Marilyn. *Prehosp Disaster Med*, 12(3), 189-194.

[25] Wylie, T., Cheanvechai, D., & Seaberg, D. (2000). Emergency response team: Hurricane Georges in Key West. *Prehosp Emerg Care*, 4(3), 222-226.

[26] Millin, M. G., Jenkins, J. L., & Kirsch, T. (2006). A comparative analysis of two external health care disaster responses following Hurricane Katrina. *Prehosp Emerg Care*, 10(4), 451-456.

[27] Kim, H., et al. (2010). Post-Nargis medical care: experience of a Korean Disaster Relief Team in Myanmar after the cyclone. *Eur*, 17(1), 37-41.

[28] Krol, D. M., et al. (2007). A mobile medical care approach targeting underserved populations in post-Hurricane Katrina Mississippi. *J Health Care Poor Underserved*, 18(2), 331-340.

[29] Sharma, A. J., et al. (2008). Chronic disease and related conditions at emergency treatment facilities in the New Orleans area after Hurricane Katrina. *Disaster Med Public Health Prep*, 2(1), 27-32.

[30] Rath, B., et al. (2007). Adverse health outcomes after Hurricane Katrina among children and adolescents with chronic conditions. *J Health Care Poor Underserved*, 18(2), 405-417.

[31] Bieberly, J., & Ali, J. (2008). Treatment adherence of the latently infected tuberculosis population (post-Katrina) at Wetmore TB Clinic, New Orleans, USA. *Int J Tuberc Lung Dis*, 12(10), 1134-1138.

[32] Karras, N. A., & Hemenway, C. S. (2007). Hurricane Katrina's impact on pediatric and adult patients with sickle cell disease. *J Health Care Poor Underserved*, 18(2), 382-393.

[33] Burton, A. (2006). Crisis not over for hurricane victims. *Environ Health Perspect*, 114(8), A462.

[34] Renukuntla, V. S., et al. (2009). Disaster preparedness in pediatric type 1 diabetes mellitus. *Pediatrics*, e973-e977, Epub Oct 12.

[35] Kamps, J. L., & Varela, R. E. (2010). Predictors of metabolic control in children with Type 1 diabetes: the impact of Hurricane Katrina in a prospective study. *Diabetes*, 88(3), 234-241, Epub Mar 24.

[36] Patton-Levine, J. K., Vest, J. R., & Valadez, A. M. (2007). Caregivers and families in medical special needs shelters: an experience during Hurricane Rita. *Am J Disaster Med*, 2(2), 81-86.

[37] Lupa, M., Molony, T., & Amedee, R. (2010). Hurricane Katrina and its effects on a regional cochlear implant program. *Laryngoscope*, 120(4), S210.

[38] van Aalst, J. A., et al. (2011). Natural disaster and crisis: lessons learned about cleft and craniofacial care from Hurricane Katrina and the west bank. *Cleft*, 48(6), 741-749, Epub Jan 29.

[39] Berry, S., et al. (2011). Care coordination in a medical home in post-Katrina New Orleans: lessons learned. *Matern*, 15(6), 782-793.

[40] Joseph, D. A., et al. (2007). Use of state cancer surveillance data to estimate the cancer burden in disaster-affected areas--Hurricane Katrina. *Prehosp Disaster Med*, 22(4), 282-290.

[41] Ridenour, M. L., et al. (2007). Displacement of the underserved: medical needs of Hurricane Katrina evacuees in West Virginia. *J Health Care Poor Underserved*, 18(2), 369-381.

[42] Jenkins, J. L., et al. (2009). Changes needed in the care for sheltered persons: a multistate analysis from Hurricane Katrina. *Am J Disaster Med*, 4(2), 101-106.

[43] Rami, J. S., et al. (2008). A school of nursing's experience with providing health care for Hurricane Katrina evacuees. *Abnf J*, 19(3), 102-106.

[44] Murray, K. O., et al. (2009). Emerging disease syndromic surveillance for Hurricane Katrina evacuees seeking shelter in Houston's Astrodome and Reliant Park Complex. *Public Health Rep.*, 124(3), 364-371.

[45] Brown, O. W. (2006). Using international practice techniques in Texas: Hurricane Katrina experiences: receiving patients in Longview, Texas, 350 miles from ground zero. *Pediatrics*, 117(5), Pt 3, S439-S441.

[46] Centers for Disease Control and Prevention. (1998). Acute hemorrhagic conjunctivitis--St. Croix, U.S. Virgin Islands, September-October 1998. *MMWR Morb Mortal Wkly Rep.*, 47(42), 899-901.

[47] Centers for Disease Control and Prevention. (2005). Norovirus outbreak among evacuees from Hurricane Katrina--Houston, Texas, September. *MMWR Morb Mortal Wkly Rep.*, 54(40), 1016-1018.

[48] Rigau-Perez, J. G., et al. (2001). Dengue activity in Puerto Rico during an interepidemic period (1995-1997). *Am J Trop Med Hyg*, 64(1-2), 75-83.

[49] Beatty, M. E., et al. (2007). Mosquitoborne infections after Hurricane Jeanne, Haiti, 2004. *Emerg Infect Dis*, 13(2), 308-310.

[50] Bhunia, R., & Ghosh, S. (2011). Waterborne cholera outbreak following Cyclone Aila in Sundarban area of West Bengal, India, 2009. *Trans*, 105(4), 214-219, Epub Feb 25.

[51] Guidry, V. T., & Margolis, L. H. (2004). Unequal respiratory health risk: using GIS to explore hurricane-related flooding of schools in Eastern North Carolina. *Environ Res., 2005,* 98(3), 383-389, Epub Dec 15.

[52] Rath, B., et al. (2011). Adverse respiratory symptoms and environmental exposures among children and adolescents following Hurricane Katrina. *Public,* 126(6), 853-860.

[53] Rabito, F. A., et al. (2008). Children's respiratory health and mold levels in New Orleans after Katrina: a preliminary look. *J Allergy Clin Immunol,* 121(3), 622-625, Epub Jan 7.

[54] Wu, F., Biksey, T., & Karol, M. H. (2007). Can mold contamination of homes be regulated? Lessons learned from radon and lead policies. *Environ Sci Technol,* 41(14), 4861-4867.

[55] Mielke, H. W., Gonzales, C. R., & Mielke, P. W., Jr. (2011). The continuing impact of lead dust on children's blood lead: comparison of public and private properties in New Orleans. *Environ Res,* 111(8), 1164-1172.

[56] Campanella, R., & Mielke, H. W. (2008). Human geography of New Orleans' high-lead geochemical setting. *Environ Geochem Health,* 30(6), 531-540.

[57] Mielke, H. W., et al. (2006). Hurricane Katrina's impact on New Orleans soils treated with low Lead Mississippi River alluvium. *Environ Sci Technol,* 40(24), 7623-7628.

[58] Abel, M. T., et al. (2010). Lead distributions and risks in New Orleans following Hurricanes Katrina and Rita. *Environ Toxicol Chem,* 29(7), 1429-1437.

[59] Rabito, F. A., et al. (2011). Environmental lead after Hurricane Katrina: implications for future populations. *Environ, 2012,* 120(2), 180-184, Epub Nov 3.

[60] Presley, S. M., et al. (2010). Metal concentrations in schoolyard soils from New Orleans, Louisiana before and after Hurricanes Katrina and Rita. *Chemosphere,* 67-73, Epub Apr 10.

[61] Zahran, S., et al. (2010). New Orleans before and after Hurricanes Katrina/ Rita: a quasi-experiment of the association between soil lead and children's blood lead. *Environ,* 44(12), 4433-4440.

[62] Van Sickle, D., et al. (2007). Carbon monoxide poisoning in Florida during the 2004 hurricane season. *Am J Prev Med,* 32(4), 340-346.

[63] Centers for Diseaes Control and Prevention. (2005). Carbon monoxide poisonings after two major hurricanes--Alabama and Texas, August-October. *MMWR Morb Mortal Wkly Rep. 2006,* 55(9), 236-239.

[64] Centers for Disease Control and Prevention. (2008). Carbon monoxide exposures after hurricane Ike- Texas, September. *MMWR Morb Mortal Wkly Rep. 2009,* 58(31), 845-849.

[65] Fife, C. E., et al. (2009). Dying to play video games: carbon monoxide poisoning from electrical generators used after hurricane Ike. *Pediatrics,* 123(6), e1035-e1038.

[66] Harduar-Morano, L., & Watkins, S. (2011). Review of unintentional non-fire-related carbon monoxide poisoning morbidity and mortality in Florida, 1999-2007. *Public,* 126(2), 240-250.

[67] Gallagher, J. J., et al. (2006). Can burn centers evacuate in response to disasters? *J Burn Care Res,* 27(5), 596-599.

[68] Seybold, U., et al. (2007). Colonization with multidrug-resistant organisms in evacuees after Hurricane Katrina. *Infect Control Hosp Epidemiol,* 28(6), 726-729, Epub Apr 20.

[69] Baldwin, S., et al. (2006). Moving hospitalized children all over the southeast: interstate transfer of pediatric patients during Hurricane Katrina. *Pediatrics,* 117(5), Pt 3, S416-20.

[70] Lowe, C. G. (2009). Pediatric and neonatal interfacility transport medicine after mass casualty incidents. *J Trauma,* 67(2 Suppl), S168-71.

[71] Barkemeyer, B. M. (2006). Practicing neonatology in a blackout: the University Hospital NICU in the midst of Hurricane Katrina: caring for children without power or water. *Pediatrics,* 117(5), Pt 3, S369-74.

[72] Cone, D. C., & Cummings, B. A. (2006). Hospital disaster staffing: if you call, will they come? *Am J Disaster Med,* 1(1), 28-36.

[73] Moll, D. M., et al. (2007). Health impact of water and sanitation infrastructure reconstruction programmes in eight Central American communities affected by Hurricane Mitch. *J Water Health,* 5(1), 51-65.

[74] Auceda, R. (1999). A land of possibility: Honduras' Mosquito Coast. *Perspect Health,* 4(2), 8-11.

[75] Needle, S. (2008). Pediatric private practice after Hurricane Katrina: proposal for recovery. *Pediatrics,* 122(4), 836-842.

[76] Boom, J. A., Dragsbaek, A. C., & Nelson, C. S. (2007). The success of an immunization information system in the wake of Hurricane Katrina. *Pediatrics,* 119(6), 1213-1217.

[77] Urquhart, G. A., et al. (2007). Immunization information systems use during a public health emergency in the United States. *J Public Health Manag Pract,* 13(5), 481-485.

[78] Barrios, R. E., et al. (2000). Nutritional status of children under 5 years of age in three hurricane-affected areas of Honduras. *Rev Panam Salud Publica*, 8(6), 380-384.

[79] Ponnapakkam, A., & Gensure, R. (2008). Effects of stress after hurricanes katrina and rita on pubertal disorders in children. *Ochsner J*, 8(3), 129-133.

[80] Zahran, S., et al. (2010). Maternal hurricane exposure and fetal distress risk. *Risk Anal*, 30(10), 1590-1601.

[81] Kinney, D. K., et al. (2008). Autism prevalence following prenatal exposure to hurricanes and tropical storms in Louisiana. *J Autism Dev Disord*, 38(3), 481-488, Epub 2007 Jul 6.

[82] Harville, E., Xiong, X., & Buekens, P. (2010). Disasters and perinatal health:a systematic review. *Obstet*, 65(11), 713-728.

[83] Quinn, D., et al. (2008). Addressing concerns of pregnant and lactating women after the 2005 hurricanes: the OTIS response. *MCN Am J Matern Child Nurs*, 33(4), 235-241.

[84] Lobato, M. N., et al. (2007). Impact of Hurricane Katrina on newborn screening in Louisiana. *Pediatrics*, e749-e755.

[85] Russoniello, C. V., et al. (2002). Childhood posttraumatic stress disorder and efforts to cope after Hurricane Floyd. *Behav Med*, 28(2), 61-71.

[86] Shaw, J. A., et al. (1995). Psychological effects of Hurricane Andrew on an elementary school population. *J Am Acad Child Adolesc Psychiatry*, 34(9), 1185-1192.

[87] Vernberg, E. M., et al. (1996). Prediction of posttraumatic stress symptoms in children after hurricane Andrew. *J Abnorm Psychol*, 105(2), 237-248.

[88] La Greca, A., et al. (1996). Symptoms of posttraumatic stress in children after Hurricane Andrew: a prospective study. *J Consult Clin Psychol*, 64(4), 712-723.

[89] Marsee, M. A. (2008). Reactive aggression and posttraumatic stress in adolescents affected by Hurricane Katrina. *J Clin Child Adolesc Psychol*, 37(3), 519-529.

[90] Weems, C. F., et al. (2009). Effect of a school-based test anxiety intervention in ethnic minority youth exposed to Hurricane Katrina. *Journal of Applied Developmental Psychology*, 218-226.

[91] Osofsky, H. J., et al. (2009). Posttraumatic stress symptoms in children after Hurricane Katrina: predicting the need for mental health services. *Am J Orthopsychiatry*, 79(2), 212-220.

[92] Scheeringa, M. S., & Zeanah, C. H. (2008). Reconsideration of harm's way: onsets and comorbidity patterns of disorders in preschool children and their caregivers following Hurricane Katrina. *J Clin Child Adolesc Psychol,* 37(3), 508-518.

[93] Spell, A. W., et al. (2008). The moderating effects of maternal psychopathology on children's adjustment post-Hurricane Katrina. *J Clin Child Adolesc Psychol,* 37(3), 553-563.

[94] Mc Dermott, B. M., & Palmer, L. J. (2002). Postdisaster emotional distress, depression and event-related variables: findings across child and adolescent developmental stages. *Aust N Z J Psychiatry,* 36(6), 754-761.

[95] Shannon, M. P., et al. (1994). Children exposed to disaster: I. Epidemiology of post-traumatic symptoms and symptom profiles. *J Am Acad Child Adolesc Psychiatry,* 33(1), 80-93.

[96] Kronenberg, M. E., et al. (2010). Children of Katrina: lessons learned about postdisaster symptoms and recovery patterns. *Child Dev,* 81(4), 1241-1259.

[97] Kar, N., & Bastia, B. K. (2006). Post-traumatic stress disorder, depression and generalised anxiety disorder in adolescents after a natural disaster: a study of comorbidity. *Clin Pract Epidemiol Ment Health,* 2(17), 17.

[98] Garrison, C. Z., et al. (1995). Posttraumatic stress disorder in adolescents after Hurricane Andrew. *J Am Acad Child Adolesc Psychiatry,* 34(9), 1193-1201.

[99] Kilpatrick, D. G., et al. (2003). Violence and risk of PTSD, major depression, substance abuse/dependence, and comorbidity: results from the National Survey of Adolescents. *J Consult Clin Psychol,* 71(4), 692-700.

[100] Lonigan, C. J., et al. (1994). Children exposed to disaster: II. Risk factors for the development of post-traumatic symptomatology. *J Am Acad Child Adolesc Psychiatry,* 33(1), 94-105.

[101] La Greca, A. M., Silverman, W. K., & Wasserstein, S. B. (1998). Children's predisaster functioning as a predictor of posttraumatic stress following Hurricane Andrew. *J Consult Clin Psychol,* 66(6), 883-892.

[102] Weems, C. F., et al. (2007). Predisaster trait anxiety and negative affect predict posttraumatic stress in youths after Hurricane Katrina. *J Consult Clin Psychol,* 75(1), 154-159.

[103] Olteanu, A., et al. (2011). Persistence of mental health needs among children affected by Hurricane Katrina in New Orleans. *Prehosp Disaster Med,* 26(1), 3-6.

[104] Mc Laughlin, K. A., et al. (2009). Serious emotional disturbance among youths exposed to Hurricane Katrina 2 years postdisaster. *J Am Acad Child Adolesc Psychiatry*, 48(11), 1069-1078.

[105] Bloom, B., & Cohen, R. A. (2007). Summary health statistics for U.S. children: National Health Interview Survey, 2006. *Vital Health Stat 10*, 10(234), 1-79.

[106] Mc Laughlin, K. A., et al. (2010). Trends in serious emotional disturbance among youths exposed to Hurricane Katrina. *J Am Acad Child Adolesc Psychiatry*, 49(10), 990-1000, 1000e1-2.

[107] Green, B. L., et al. (1991). Children and disaster: age, gender, and parental effects on PTSD symptoms. *J Am Acad Child Adolesc Psychiatry*, 30(6), 945-951.

[108] Allen, B., et al. (2010). Perceptions of psychological first aid among providers responding to Hurricanes Gustav and Ike. *J*, 23(4), 509-513.

[109] Caffo, E., & Belaise, C. (2003). Psychological aspects of traumatic injury in children and adolescents. *Child Adolesc Psychiatr Clin N Am*, 12(3), 493-535.

[110] Salloum, A, & Overstreet, S. (2012). Grief and trauma intervention for children after disaster: exploring coping skills versus trauma narration. *Behav*, 50(3), Epub 2012 Jan 12, 169-179.

[111] Scheeringa, M. S., et al. (2007). Feasibility and effectiveness of cognitive-behavioral therapy for posttraumatic stress disorder in preschool children: two case reports. *J Trauma Stress*, 20(4), 631-636.

[112] Chemtob, C. M., Nakashima, J. P., & Hamada, R. S. (2002). Psychosocial intervention for postdisaster trauma symptoms in elementary school children: a controlled community field study. *Arch Pediatr Adolesc Med*, 156(3), 211-216.

[113] Cohen, J. A., et al. (2009). Treating traumatized children after Hurricane Katrina: Project Fleur-de lis. *Clin Child Fam Psychol Rev*, 12(1), 55-64.

[114] Jaycox, L. H., et al. (2010). Children's mental health care following Hurricane Katrina: a field trial of trauma-focused psychotherapies. *J*, 23(2), 223-231.

[115] Garrett, A. L., et al. (2007). Children and megadisasters: lessons learned in the new millennium. *Adv Pediatr*, 54, 189-214.

[116] Mc Laughlin, K. A., et al. (2009). Serious emotional disturbance among youths exposed to Hurricane Katrina 2 years postdisaster. *J Am Acad Child Adolesc Psychiatry*, 48(11), 1069-1078.

[117] Moore, K. W., & Varela, R. E. (2010). Correlates of long-term posttraumatic stress symptoms in children following Hurricane Katrina. *Child*, 41(2), 239-250.

[118] Schoenbaum, M., et al. (2009). Promoting mental health recovery after hurricanes Katrina and Rita: what can be done at what cost. *Arch Gen Psychiatry*, 66(8), 906-914.

Chapter 9

The Impact of Hurricane Debbie (1961) and Hurricane Charley (1986) on Ireland

Kieran R. Hickey and Christina Connolly-Johnston

Additional information is available at the end of the chapter

http://dx.doi.org/10.5772/54039

1. Introduction

Ireland has long been in receipt of the tail-end of a small number of Atlantic hurricanes and is the most affected country in Europe by these storms. Although almost never actual hurricanes by the time they reach Ireland, they have still caused loss of life, injuries and extensive damage through high winds, heavy rainfall and storm surges along the coast. Little research has been carried out on these events until recently and their effects have been subsumed into the general mid-latitude storm record [1].

The chapter will investigate and analyse in detail the impact of the worst two hurricane events or tail-ends to affect Ireland over the last 50 years as both storms caused fatalities and injuries and extensive damage across the island of Ireland. In addition Hurricane Debbie in 1961 is the only hurricane to have ever made landfall in Ireland (as a Category 1 event) that is known about, although ongoing research into the historical records of North Atlantic may reveal others, especially in the pre-weather satellite era. Hurricane Charley in 1986 was the tail-end of a hurricane when it hit Ireland, but was considered an extra tropical storm no when it passed by the south coast of Ireland.

In detail this chapter will provide a detailed assessment and analysis of these two events. Firstly it will assess the generation and tracks of the hurricanes. This will be done in the context of Hickey's assessment [2]. The meteorological effects of the two events will be systematically analysed as they passed over the country using hourly data from the relevant meteorological stations around the island of Ireland. The impact of the two events will be assessed using a wide variety of sources including local, regional and national media, local and national government records amongst others. This approach will also help to provide the first meaningful estimates of the financial cost of these two events on Ireland.

2. Data sources

Meteorological information was derived from existing meteorological stations in Ireland as well as satellite imagery of Hurricane Charley in 1986. No such imagery exists for Hurricane Debbie in 1961. Hourly wind data and other parameters were analysed by event.

The information on the impact of events were gleaned from a wide variety of local and national newspaper sources throughout Ireland with special emphasis on the local newspapers in the worst affected areas of the country. As a result of the number of fatalities and the scale of damage both events generated considerable media coverage. The information from these sources was analysed according to type of damage and the exact location where they occurred. The newspapers consulted included the Anglo-Celt, Connacht Tribune, Connaught Telegraph, Irish Independent, Irish Press, Munster Express, Sunday Independent, The Kerryman, Western People and Westmeath Examiner.

3. Meteorological analysis

3.1. Origin and track

Hurricane Debbie initially formed as a storm west of Africa on the 7th September 1961 and immediately started moving westwards and intensifying and was given hurricane status on the 11th of September and quickly reached Category 3 intensity with maximum wind speeds of 195 km/h. On the 15th of September the hurricane turned northwards off Cape Verde Islands and then headed northeastwards heading towards Ireland and Europe. It made landfall in Dooega on Achill Island, Co. Mayo off the west coast of Ireland on the 16th of September but passed back out into the Atlantic before tracking along the coast of Scotland and then Norway and finally dissipating over Russia (Fig 1). This hurricane had no effects on the eastern side of the Atlantic which is somewhat unusual but is considered a major contributor to the plane crash on the Cape Verde Islands which cost the lives of 60 people [3,4].

Hurricane Charley is quite different. This hurricane formed in the southeastern Gulf of Mexico on the 13th of August then headed northeastwards through Georgia and South Carolina, USA. It then turned northwards briefly travelling parallel to the eastern seaboard of the USA before again tracking northeastwards across the Atlantic towards Ireland and Europe. The hurricane did not exceed Category 1 status with maximum wind speeds of 130km/h and lowest central pressure of 987hPa. This depression was no longer at hurricane strength when it crossed the southern third of Ireland across Wales and the midlands of the UK before finally dissipating in the North Sea near Denmark on the 25th of August (Fig. 2). In the USA this hurricane caused 5 deaths including 3 in a plane crash and over $15 million in damages [5].

http://dx.doi.org/10.5772/54039

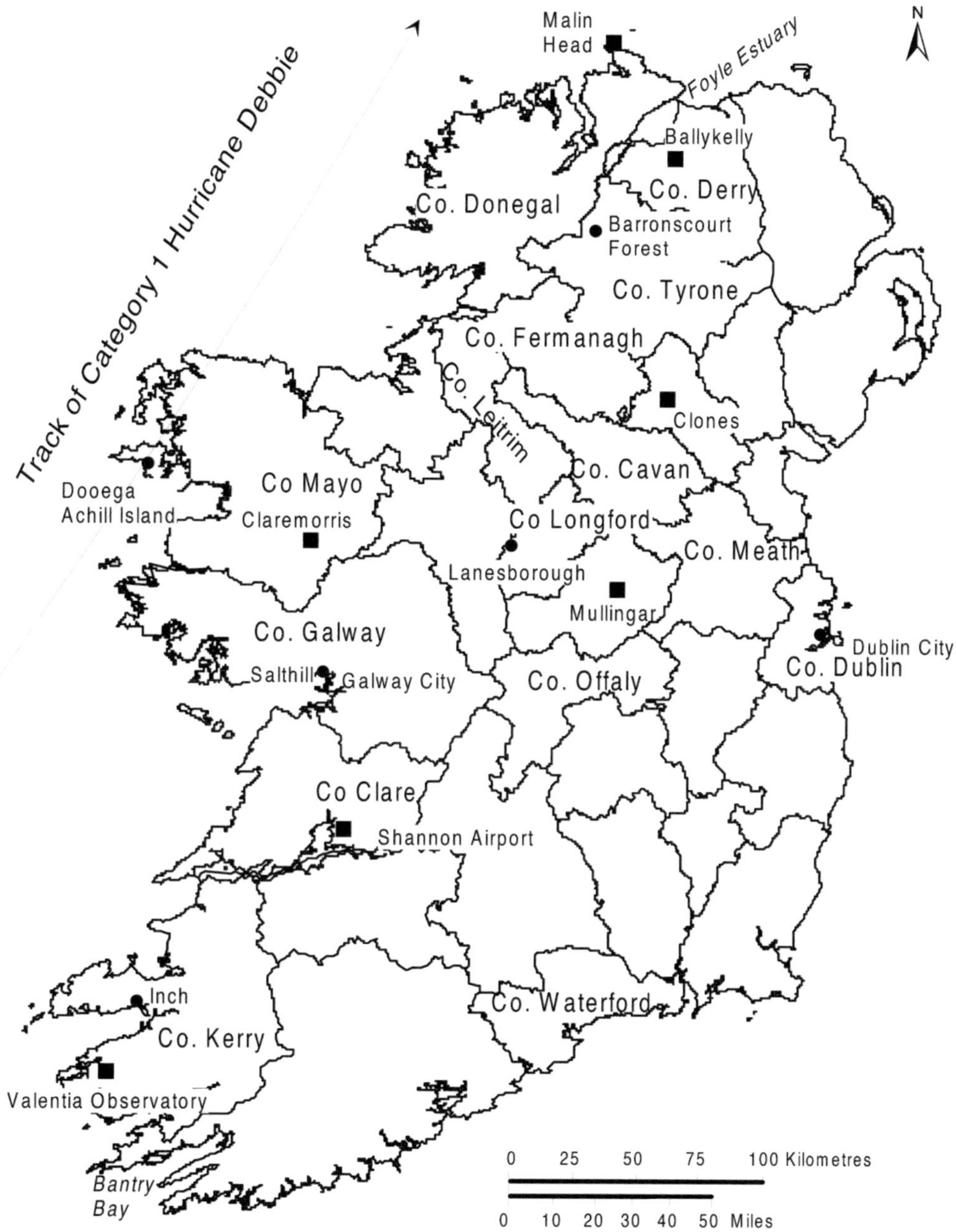

Figure 1. The track of Hurricane Debbie over Ireland, September 16th, 1961

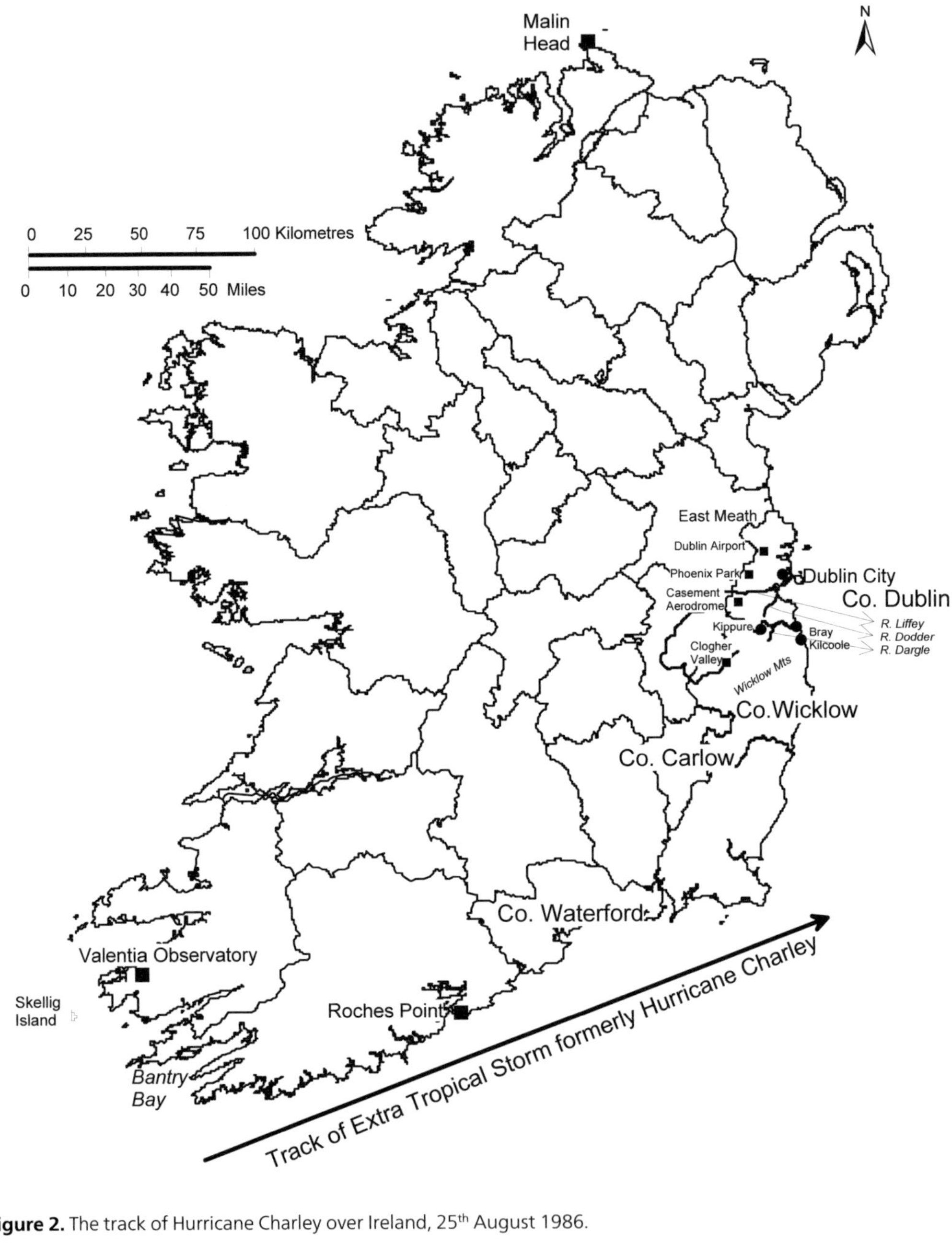

Figure 2. The track of Hurricane Charley over Ireland, 25th August 1986.

3.2. Barometric pressure

Since 1960 Hickey [6] identified that the lowest Western European barometric pressure reading for any of the tail-end of hurricane events in the survey is 950 hPa which was recorded between Ireland and Scotland and this came from Hurricane Debbie. This was the lowest pressure recorded for this hurricane itself despite the fact that by this stage it was a Category 1 at best. On land the lowest pressures recorded for Debbie were in Co. Mayo at 963 hPa [7]. Hurricane Charley in Ireland was not associated with any significantly low barometric pressure readings.

4. Wind

The highest recorded value of any of the events in this survey was from Hurricane Debbie in 1961. At Malin Head on the extreme NW tip of Ireland a gust of 182km/h was recorded. Other exceptional gusts were recorded at Shannon Airport, Rep. of Ireland at 172km/h, Ballykelly, Northern Ireland at 171km/h, Tiree, Scotland and Snaefell, Isle of Man both at 167km/h, Clones, Rep. of Ireland at 161km/h, Kyle of Lochalsh, Scotland 159km/h and Mullingar, Rep. of Ireland with 146km/h [8,9]. All the above mentioned stations in the Rep. of Ireland and Malin Head and Valentia Observatory were and still are all time record gusts. In addition maximum 10 minute mean wind speed records were set for meteorological stations at Claremorris, Mullingar and Shannon Airport and still stand. These records indicate the exceptional strength of the winds associated with Hurricane Debbie.

In terms of sustained winds the values are obviously lower than that of the gusts but equally important in terms of generating all sorts of damage from the coast moving inland. Unsurprisingly Hurricane Debbie in 1961 also generated the highest sustained values identified [10,11] and the detailed hourly wind values for both Valentia Observatory, SW Ireland and Malin Head, NW Ireland can be outlined. Fig. 3 shows the rising values of sustained hourly wind speed as the hurricane travels offshore along the west coast of Ireland t the two aforementioned meteorological stations.

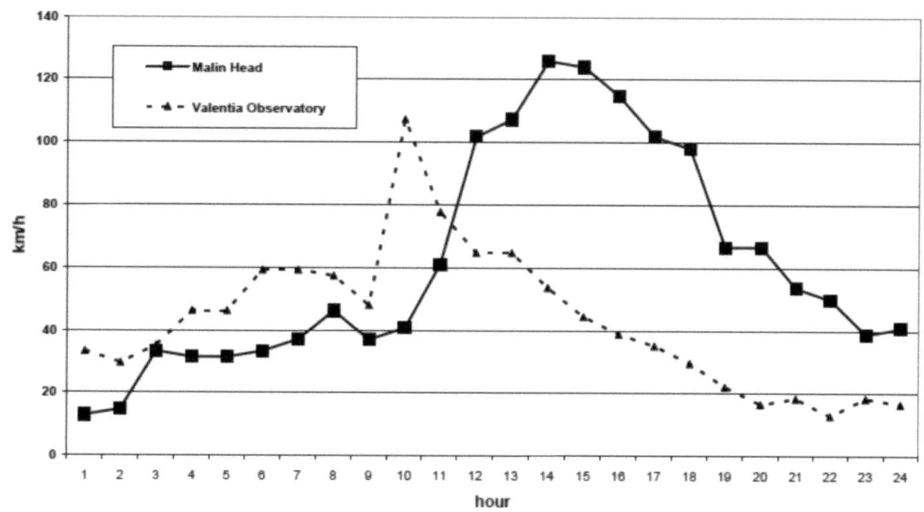

Figure 3. Comparison of the Sustained Hourly Wind Speed of Hurricane Debbie 16th September 1961 for Valentia Observatory and Malin Head Meteorological Stations

Values at Valentia Observatory are higher than that of Malin Head until noon on the 16th September 1961. Initial values at Valentia Observatory did not exceed 60km/h up to and including the 9am reading, however the maximum value of 107km/h was recorded at 10am showing a massive elevation of wind speed in a very short time period. For the rest of the day the values of wind speed gradually decline at Valentia Observatory. From 10am onwards wind speeds rapidly rose at Malin Head and reached their maximum value at 2pm with a sustained value of 126km/h, an hour later the wind speed had barely dropped to 124km/h, thereafter as the hurricane moved away the wind speeds started to decline but even at 4pm and 5pm the wind speed was above 100km per hour unlike Valentia Observatory where the peak was relatively brief and at 6pm the wind had just dropped below 100km/h, thereafter the wind speed dropped more rapidly and by midnight was hovering around 40km/h, still twice the values being recorded at Valentia Observatory.

The sustained hourly wind data from 25th of August 1986 for Hurricane Charley shows a number of small peaks as well as the main one at Valentia Observatory. Small peaks of wind strength at or above 40km/h occur at 4am, 9 to 11am and 3pm to 5pm (Fig. 4). The main peak with winds between 50 and 65km/h occurred between 7pm and 11pm thereafter wind speed starts to diminish. Much higher gusts were recorded at Brixham in Devon and Gwennap Head in Cornwall, both in England recording 121km/h and 114km/h respectively. Many southern Irish meteorological stations recorded gusts of between 90km/h and 102km/h but not as strong as England [12]. However the effects of the passing of the hurricane did not cause any significant increase in wind speed in more northerly stations like Malin Head and as such is not included in Fig. 4.

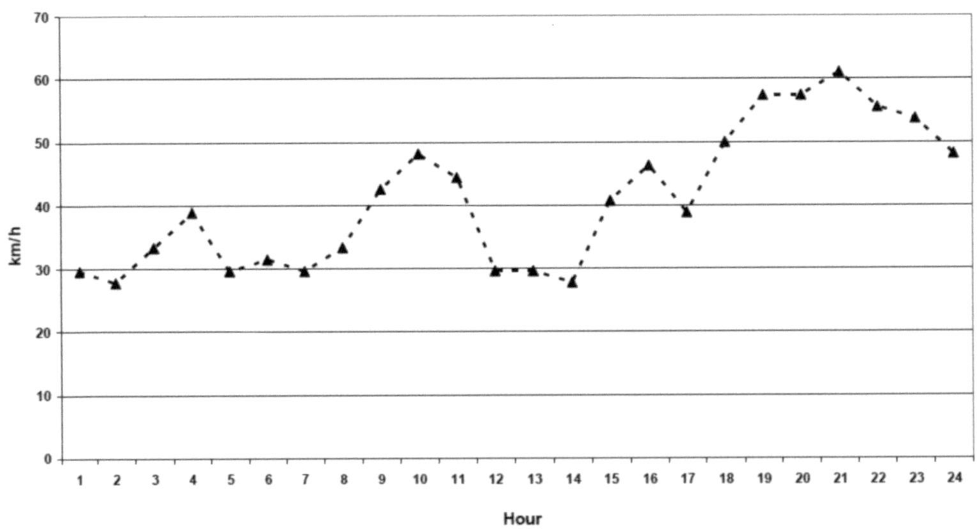

Figure 4. Sustained Hourly Wind Speed of the tail-end of Hurricane Charley on 25th of August 1986 from Valentia Observatory

5. Precipitation

Hurricane Debbie's remnants were responsible for flooding in Ireland, Scotland and Wales but the rainfall amounts although high were by no means exceptional or record breaking [13]. Some of this flooding was associated with the large lakes in the west of Ireland. Fig. 5 shows that there were two main pulses of rainfall at Valentia Observatory coming towards the end of the 16th and the start of the 17th September 1961 peaking at 9mm in an hour at 9pm on the 16th. The hourly rainfall at Malin Head shows the second lesser peak at the start of the 17th August but very limited rainfall receipt either before, during or after the hurricane, despite being closer to the landfall site of the hurricane itself.

Hurricane Charley in 1986 produced record high precipitation values for Ireland. In the mountains south of Dublin at Kippure (754m altitude) in excess of 280mm was recorded over a 24 hour period, which set a new one day record at altitude for Irish rainfall. However, because of timing errors on the data logger this figure of 280mm is viewed as a conservative estimate, the true 24 hour value could be as high as 350mm [14]. In addition, the low-lying station at Kilcoole, Co. Wicklow recorded 200mm of rainfall, setting a new Irish record for a one day total at low altitude. Additional very high daily rainfall totals were recorded at a number of other meteorological stations including the Phoenix Park, Dublin city (85.1mm), Roches Point, Co. Cork (84.4mm) Casement Aerodrome, Co. Dublin (72.4mm) and a number of others with values ranging from 50 to 65mm [15]. This amount of rainfall unsurprisingly caused significant flooding in Dublin city and also in Bray Co. Wicklow. In parts of Bray water heights reached 2.4m and over 450 buildings were affected by the flooding.

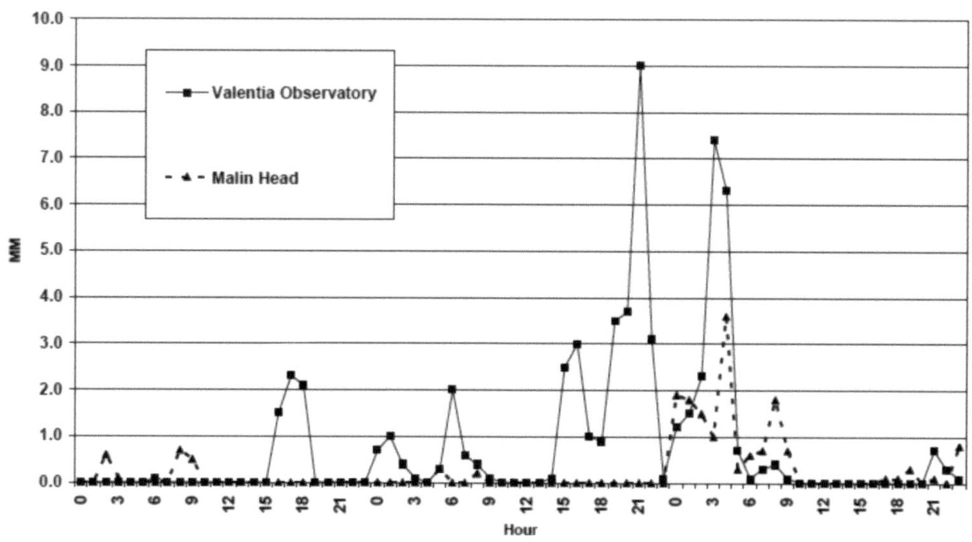

Figure 5. Hourly Rainfall at Valentia Observatory and Malin Head from Hurricane Debbie 15th-17th September 1961

The hourly rainfall from Valentia Observatory, Dublin Airport and Casement Aerodrome show the heavy rainfall associated with this event (Fig. 6). Valentia Observatory on the west coast records the passage of the event with a midday peak of hourly rainfall receipt on the 25th of August. The two stations on the east coast of Ireland closest to the major flooding record peak hourly rainfall from 4pm to midnight on the 25th of August with declining rainfall levels from then on apart from a small peak at Dublin Airport at 8am, arguably unrelated to the passage of the tail-end of the hurricane. Casement Aerodrome recorded 9.3mm in an hour at 6pm on the 25th of August the maximum hourly value recorded for any of the three stations. This is the station closest to the locations where the highest daily rainfall totals were recorded, which were in the Wicklow Mts. Valentia Observatory had a maximum value of 7.9mm at 1pm on the 25th and Dublin Airport had a maximum value of 6.8mm at 7pm on the 25th of August 1986.

When the remnants of Hurricane Charley moved on to Wales it also produced excessive rainfall but not quite as high as Dublin. At Aber in Gwynedd in Wales 135mm was recorded over an 18 hour period. Very high daily values were also recorded in England with Walshaw Dean near Halifax having 121mm of rainfall and Loggerheads near Wrexham having 109mm of rainfall. Many other locations had values near or just exceeding 100mm in a day [16].

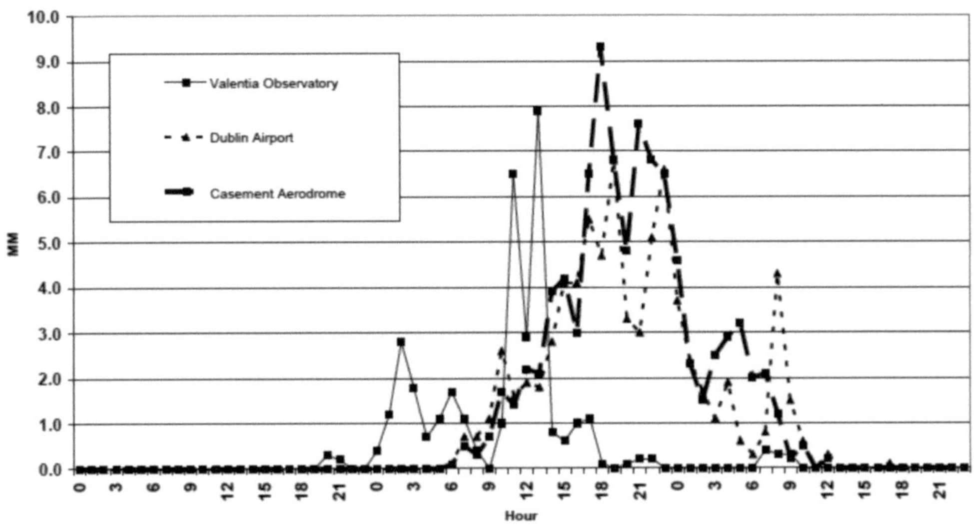

Figure 6. Hourly Rainfall at Valentia Observatory, Dublin Airport and Casement Aerodrome from the tail-end of Hurricane Charley 24th-26th August 1986

Both events are very different in terms of their origin and track yet both end up crossing Ireland either as a hurricane or extra tropical storm. Both are also significant in an Irish context but for very different reasons, Debbie for record high winds and Charley for record high precipitation. These differences have a big effect on the impacts generated by the two events as will be detailed in the next section.

6. Impact analysis

Although the type of impact varies between the two events both produced significant numbers of fatalities and injuries along with considerable economic damage.

7. Fatalities and injuries

Aside from the loss of 60 lives in a plane crash on the Cape Verde Islands Hurricane Debbie is directly responsible for the deaths of at least 17 people in Ireland. A systematic search of local and regional newspapers across Ireland was used to generate this figure which is higher than any previous published number [17,18,19]. In the worst incident four members of the one family including a one week old being taken home from hospital were killed in Co. Cavan when a tree fell on their car (Fig.1). Tree falls or driving into fallen trees were responsible for another eight deaths in eight separate incidents, three of which occurred in Co. Tyrone, Northern Ireland with one each in Cos. Donegal, Dublin, Longford, Meath and Offaly all in the Rep. of Ireland. Three fatalities occurred as a result of collapsing walls and roofs, one each in Co. Fermanagh in Northern Ireland and Dublin city and Co. Meath in the Rep. of Ireland. In addition a five year old lost his life when he was blown into a stream in Co. Fermanagh, Northern Ireland and another person drowned as a small boat capsized in Co. Derry. A large number of people suffered injuries of which at least seven of these were considered serious. The main causes of these injuries were to do with fallen trees, roofs and walls and flying debris such as roof slates.

As well as the five fatalities in the USA the weather remnants of Hurricane Charley in August 1986 caused an additional 11 fatalities in Europe, six in Ireland and five in England. Most of these fatalities were as a result of drowning in flooded rivers or being trapped in flood waters. Two of the Irish fatalities resulted from drowning incidents in the Dublin area and one of the Irish fatalities occurred as a result of drowning on a canoeing trip in Co. Carlow, Rep. of Ireland (Fig 2). At least one other death is confirmed in the newspapers but with little detail and details on the other two fatalities remain very sketchy at best. The figure of six fatalities for this event has been in all the limited published literature and it is surprising that only three or four definitive fatalities can be identified for this event using modern search methods. The five fatalities in England also resulted mainly from drowning incidents. Very few people sustained injuries in either Ireland or England.

8. Economic damage

8.1. Hurricane Debbie

Virtually no area on the island of Ireland was unaffected by the impact of Hurricane Debbie, but the worst affected areas were Cos. Mayo and Galway (Fig 1). Tens of thousands of houses and other structures including churches suffered significant structural damage varying from complete destruction to roof loss to minor damage including broken windows. Much damage was done to walls, sheds and other infrastructure. Caravan parks along the west coast were particularly badly affected with some caravans being moved up to 150 metres and many others being completely destroyed. Roads and other transport routes were blocked as a result of falling trees and electricity and telephone cables. There were extensive power outages with some areas being without power for up to four days as well as extensive disruption to transport including shipping on the Irish Sea.

There was significant damage done to the agricultural sector with the hurricane coming at one of the busiest times of the year. There was widespread damage to barns, sheds and outhouses and other agricultural buildings. In addition unharvested wheat and oats were beaten down in the fields where they stood to such an extent that it is estimated that up to a third of the crop had been lost nationally. Part of the hay crop was blown away if harvested and still uncollected in the fields. The severity of the wind can be seen by the fact that as far as 20km inland all plant life withered and died in a matter of minutes as sea spray laden with salt was carried landward by the wind. A similar effect was recorded for the 6th January 1839 storm [20]. There was extensive damage to forestry all across the Ireland. Total losses were of the order of 2% of all trees in commercial plantations on the island of Ireland but with some forests these losses were up to 24% [21]. The worst affected plantation was Baronscourt Forest in Co. Tyrone where around 300,000 mature trees were blown down. Other major losses occurred in Cos. Derry, Fermanagh, Leitrim, Galway and Clare. The force of the wind damaged many native trees throughout the country by removal of branches, leaves and shoots. This damage included a large number of sweet cherry trees at a farm in Co. Waterford [22]. Fallen trees and blown debris also killed many farm animals including cattle.

There was extensive damage to shipping along the south, west and north coasts especially at Bantry Bay, Co. Cork and the Foyle Estuary between Cos. Donegal and Derry. A very rare storm-induced tidal bore was recorded as having taken place on the Shannon river near Lanesboro, Co. Longford when the level of the river rose by 1.35m as the hurricane winds blew water upstream. This reverse flow carried many small boats upstream and onto the river banks leaving them high and dry when the wind changed direction and the river dropped almost equally as suddenly.

Very little coastal flooding was reported with the exception of Salthill, Galway city which experienced early morning severe tidal flooding. There is some evidence of small scale coastal erosion in parts of Ireland including some significant shoreline recession at Inch, Co. Kerry. However, most of the coasts were not affected due to the nature of the winds being directed obliquely or offshore and the fact that the peak winds did not coincide with high tide explains this outcome [23]. On the Lancashire coast of England and the Isle of Man extensive sand storms were recorded with significant deposits of sand inland but there are no records of this occurring in Ireland [24].

It is quite difficult to estimate the cost of Hurricane Debbie but based on the few figures that are available and the scale of the structural damage to properties and infrastructure, the widespread devastation of forestry plantations and the damage to the agricultural sector a figure in 1961 terms of between $40 million and $50 million would not be unrealistic.

8.2. Hurricane Charley

Hurricane Charley mostly affected the southern half of Ireland especially on the east coast with Cos. Dublin and Wicklow being particularly affected (Fig. 2). However, unlike Debbie most of the damage associated with this event was to do with flooding as a result of the exceptional rainfall. Particularly badly affected were parts of south Dublin city with 400 houses flooded and Bray Co. Wicklow where over a thousand residents had to be evacuated as flood waters breached the banks of the River Dargle [25]. Roads, bridges and property were extensively flooded throughout much of the southern and southeastern part of Ireland. Additional flooding was caused by the River Dodder in Dublin city bursting its banks causing significant damage mostly to private dwellings. The Dodder rises in the Dublin Mountains at an altitude of 751m OD and as such is vulnerable to flash flooding caused by high mountain rainfall as was the case during Hurricane Charley [26]. The return period of this flood event on the Dodder was estimated to be in excess of 100 years [27]. This was partly based on the recorded maximum output of the Lower Bohernabreena Reservoir on the river. 91.1 (m3/s) were recorded the third largest of the records dating back to 1886. The 13th of October 1891 value of 92.2 (m3/s) and the 28th August 1905 of 107.8 (m3/s) both events caused very significant flooding, although the areas along the Dodder at this time were much less built-up [28].

Fortunately because of the effects of the River Liffey reservoirs and their effective management, most of Dublin city was unaffected by flooding. Without the reservoirs and the flood management severe flooding would have occurred in Dublin as evidenced by the estimates of peak flow with and estimated without the reservoirs and associated dams. The peak flow downstream of the Leixlip Dam during Hurricane Charley was 170 (m^3/s) without these structures it was estimated that it discharge would reach 400 (m^3/s) producing a very different outcome in Dublin city as a result [29]. Parts of East Meath were also considerably affected by flooding, extending the worst flood-affected areas westwards from Dublin city and county [30].

The torrential rain from Charley also triggered three very small scale landslides and one slightly larger one in the Cloghoge Valley in the Wicklow Mts., felling trees and stripping the bark from those left standing as the debris slides tracked through forested areas. The three very small slides involved movement of between 310 and 533 m^3 of material whereas the larger slide involved the movement of an estimated 6,578 m^3 and covered an area of 12.75km^2. This landslide although much bigger than the other three still only represented 0.14% of the total catchment, the other three barely reaching 0.01% of the catchment [31].

Wind damage was also recorded across this area with significant damage to power and phone lines partly due to fallen trees. Initially nearly 250,000 people were affected by power failures although this was reduced down to 9,000 after two days of intensive effort. There was extensive disruption to transport both public and private across the country and to ferries in the Irish Sea. Social and sporting fixtures were also cancelled.

Once again the agricultural sector was severely affected by a combination of the heavy rain, leading to flooding and high winds. Cereal production was particularly badly affected with losses of 50% estimated for Co. Waterford alone. In addition feed, hay and silage supplies also suffered major losses so much so that in many areas in the southern half of the country the early winter housing of livestock had to be implemented as the waterlogged and sodden land could neither provide nor sustain the needs of the animals. Consequentially milk supply and farmers incomes also suffered.

The height of the tail-end of Hurricane Charley also coincided with two high tides generating localized coastal flooding and causing problems for shipping as the pounding waves threw a luxury yacht onto the heavily flooded roadway at Bray seafront in Co. Wicklow. Valentia lifeboat was called out at the peak of the storm to a distress call by a yacht off the Skellig Islands, Co. Kerry.

Again it is impossible to put an exact cost on the damage and destruction caused by the tail-end of Hurricane Charley, however some justifiable estimate can be generated by using what few figures are available in 1986 values. Around $40 million in insurance claims mostly for flooding were made to various insurance companies in Ireland [32]. The Irish Government allocated $8.65 million just for road and bridge repairs. However these values do not include the significant losses suffered by the farming sector and also uninsured losses. A true value in excess of $100 million and even as high as $125 million would not be unjustified for Ireland alone not too mention across the other affected countries in Europe.

9. Discussion and conclusions

One of the key results that has emerged from the detailed examination of the impact of Hurricane Debbie and the tail-end of Hurricane Charley is that each had its own unique character. Debbie was very much associated with wind damage and set new wind records for Ireland, some of which still stand today whereas Charley was very much a flood damage event due to the exceptional and again record breaking rainfall. In many respects the two events represent two extremes of the possible types of meteorological effects of the tail-end of a hurricane on Ireland and Europe. Worryingly for Ireland would be the case where an event occurred that contained both exceptionally high winds and exceptionally high rainfall. This could have a devastating impact not just on Ireland but probably on other parts of Europe as well.

However, it is very hard to consider how likely this perfect tail-end/hurricane is to occur for a number of reasons. Firstly, not enough research has been carried out on past Irish storm records, in particular there is a need to focus on unusually severe storms that affected Ireland over the last several hundred years during the months of August, September and October and which may have been mistakenly identified as early mid-latitude storms and not ones of tropical origin. Irish storm records go back to the first millennium AD due to the survival of a number of monastic Irish Annals covering this time period up to the middle of the second millennium. These annals record significant weather events including major devastating storms, a few of which might be hurricane in origin.

Secondly, even if potential tail-end of hurricanes or hurricanes themselves were identified, there would be enormous difficulties in categorically proving their tropical origin or at least producing enough evidence to suggest that this was even likely, and in many cases no definitive proof would ever be found particularly events that would predate AD 1800 and especially AD 1700.

Consideration must also be given to rising sea-surface temperatures in the tropics and how this will gradually enlarge the areas lattitudinally where hurricanes could potentially form. This has huge implications for the potential loss of life and damage in tropical areas and this in turn will also have a potentially significant impact on the frequency of the tail-ends of hurricanes and hurricanes themselves reaching and affecting Ireland and Europe. In addition their intensity may be increased as well leading to greater damage and destruction and the potential for more loss of life and injuries. As a result more attention needs to be paid to these events and the frequency with which they affect Western Europe. However, at present modeling of likely future hurricane activity has failed to indicate any significant increase or decrease but it is noted that much research needs to be carried out particularly in dealing with the chaotic nature of the climate system and in the response of the climate to radiative forcing in order to develop much more suitable models for prediction [33].

It is clear that the potential impact of the tail-ends of hurricanes or hurricanes themselves on both Ireland and Europe should not be underestimated as the impacts of Hurricane Debbie and the tail-end of Hurricane Charley on Ireland has shown. Loss of life and injuries can be more severe and greater than that of mid-latitude storms and the potential scale of damage and destruction can be very significant whether through wind damage or flooding or a combination of both.

More recently in 2012 the impact of tail-end of Hurricane Katia on Ireland and Europe with one fatality and extensive damage stretching from Ireland to Russia and the impact of the tail-end of Hurricane Ophelia also in 2012 which brought bad weather to Europe shows that the threat remains ever present. With rising sea surface temperatures in the tropics in the Atlantic the potential threat of more of these events reaching Ireland and Europe cannot be ignored even though there will still be clusters of years when no tail-ends will reach as far northwards [34].

Future research will focus on identifying the hurricane component of the Irish storm record and in doing so identify what contribution the tail-ends of hurricanes and even hurricanes themselves make to precipitation and wind receipt in Ireland and disentangle these much rarer events from the normal mid-latitude storm signal.

Acknowledgements

My thanks to Dr. Siubhán Comer for drawing the maps and to Met Éireann for the hourly wind and rain data.

Author details

Kieran R. Hickey and Christina Connolly-Johnston

*Address all correspondence to: kieran.hickey@nuigalway.ie

Department of Geography, National University of Ireland, Galway, Rep. of Ireland

References

[1] Hickey K.R. (2011a) The impact of hurricanes on the weather of Western Europe in Lupo A. (ed) Recent Hurricane Research – Climate, Dynamics and Societal Impacts, InTech, Croatia, p77-84

[2] Hickey K.R. (2011a) The impact of hurricanes on the weather of Western Europe in Lupo A. (ed) Recent Hurricane Research – Climate, Dynamics and Societal Impacts, InTech, Croatia, p77-84

[3] Met Éireann (2012) Exceptional Weather Events Database, Retrieved 31st May 2012.

[4] National Oceanic and Atmospheric Administration (NOAA), USA (2012) Atlantic Hurricane Database, Retrieved 6th June 2012.

[5] National Oceanic and Atmospheric Administration (NOAA), USA (2012) Atlantic Hurricane Database, Retrieved 6th June 2012.

[6] Hickey K.R. (2011a) The impact of hurricanes on the weather of Western Europe in Lupo A. (ed) Recent Hurricane Research – Climate, Dynamics and Societal Impacts, InTech, Croatia, p77-84

[7] Met Éireann (2012) Exceptional Weather Events Database, Retrieved 31st May 2012.

[8] Bedford R. (2008) Yesterday's weather, Self Published.

[9] Met Éireann (2012) Exceptional Weather Events Database, Retrieved 31st May 2012.

[10] Hickey K.R. (2011b) The hourly gale record from Valentia Observatory, SW Ireland 1874-2008 and some observations on extreme wave heights in the NE Atlantic, Climatic Change, Vol. 106 (3) p483-506.

[11] MacClenahan P., McKenna J., Cooper J.A.G. and O'Kane B. (2001). Identification of highest magnitude coastal storm events over Western Ireland on the basis of wind

http://dx.doi.org/10.5772/54039

speed and duration thresholds, International Journal of Climatology Vol. 21, p829-842.

[12] Eden P. (2008) Great British weather disasters, London, Continuum.

[13] National Oceanic and Atmospheric Administration (NOAA), USA (2012) Atlantic Hurricane Database, Retrieved 6th June 2012.

[14] Graham E. (2006) 200mm fall in Ireland, Weather, Vol.6 (5), p151.

[15] Met Éireann (1986) Storms cause widespread flooding, Monthly Weather Bulletin, August 1986.

[16] Eden P. (2008) Great British weather disasters, London, Continuum.

[17] Hickey K.R. (2011a) The impact of hurricanes on the weather of Western Europe in Lupo A. (ed) Recent Hurricane Research – Climate, Dynamics and Societal Impacts, InTech, Croatia, p77-84

[18] National Oceanic and Atmospheric Administration (NOAA), USA (2012) Atlantic Hurricane Database, Retrieved 6th June 2012.

[19] Met Éireann (2012) Exceptional Weather Events Database, Retrieved 31st May 2012.

[20] Carr P. (1992) The night of the big wind: the story of the legendary big wind of 1839, Ireland's greatest natural disaster, 2nd Edition, Belfast, White Row Press.

[21] Cruickshank J.G., Stephens N. and Symons L.J. (1962) Report of the hurricane in Ireland on Saturday, 16th September 1961, Irish Naturalists Journal, Vol. 14 (1) p4-12.

[22] Kennedy N.D.G. and Kavanagh T. (1968) Bacterial canker (Pseudomonas mors-prunorum) of Sweet Cherries, Irish Journal of Agricultural Research, Vol. 7 (1), p134-136.

[23] Cooper J.A.G., Jackson D.W.T., Navas F., McKenna J. and Malvarez G. (2004) Identifying storm impacts on an embayed, high-energy coastline: some examples from western Ireland, Marine Geology, No. 210, p261-280.

[24] Bedford R. (2008) Yesterday's weather, Self Published.

[25] de Bruijn E.I.F. and Brandsma T. (2000) Rainfall prediction for a flooding event in Ireland caused by the remnants of Hurricane Charley, Journal of Hydrology No. 239, p148-161.

[26] Cawley A.M. and Cunnane C. (2005) A selection of extreme flood events – the Irish experience, Proceedings of the National Hydrology Seminar, Tullamore, Co. Offaly, p14-25.

[27] Mangan B. (1999) Flood risk assessment and communication: the Irish experience, Proceedings of the First RIPARIUS Workshop, Brussels, 27-28 October 1998, p75-86.

[28] MacDonald D.E. and Molyneux J.D. (2002) Rehabilitation of the Upper and Lower Bohernabreena spillways in Tedd P. (ed) Reservoirs in a Changing World, Proceed-

ings of the 12th Conference of the British Dam Society, Dublin, 4-8 September 2002, p274-285.

[29] Fitzpatrick J. and Bree T. (2001) Flood risk management through reservoir storage and flow control, Proceedings of the National Hydrology Seminar, Tullamore, Co. Offaly, p87-96.

[30] Bhattarani K. and Baigent S. (2009) The hydrological analysis for the final Fingal East Meath flod risk assessment and management study, Proceedings of the Irish National Hydrology Conference, Tullamore, Co. Offaly, p58-67.

[31] Bourke M.C. and Thorpe M. (2005) Rainfall-triggered slope failures in eastern Ireland, Irish Geography, Vol.38 (1), p1-22.

[32] de Bruijn E.I.F. and Brandsma T. (2000) Rainfall prediction for a flooding event in Ireland caused by the remnants of Hurricane Charley, Journal of Hydrology No. 239, p148-161.

[33] Villarini G. and Vecchi G.A. (2012) Twenty-first century projections of North Atlantic tropical storms from CMIP5 models, Nature Climate Change, Vol.2 p604-607.

[34] National Oceanic and Atmospheric Administration (NOAA), USA (2012) Atlantic Hurricane Database, Retrieved 6th June 2012.